D0398637

SYSTEMATIC

SYSTEMATIC

HOW SYSTEMS BIOLOGY IS TRANSFORMING MODERN MEDICINE

James R. Valcourt

Bloomsbury Sigma
An imprint of Bloomsbury Publishing Plc

1385 Broadway
New York
NY 10018
USA

50 Bedford Square
London
WC1B 3DP
UK

www.bloomsbury.com
BLOOMSBURY and the Diana logo are trademarks of
Bloomsbury Publishing Plc

First published 2017

British Library Cataloguing-in-Publication Data
A catalogue record for this book is available from the British Library.

Library of Congress Cataloguing-in-Publication Data is available.

ISBN (hardback) 978-1-63286-029-3
ISBN (ebook) 978-1-63286-031-6

2 4 6 8 10 9 7 5 3 1

Typeset in Bembo Std by RefineCatch Limited, Bungay, Suffolk
Printed and bound in Great Britain by CPI Group (UK) Ltd,
Croydon CR0 4YY

For my family, my teachers, Amy, and
the original Broadway cast of Hamilton

Author's Note

In introducing general readers to systems biology, I am including interesting research that is guided by the systems idea rather than pretending to provide a comprehensive or even fully representative overview of the field.

The challenge to responsibly communicate new ideas that may be revised or even discarded with future research is one I take very seriously. The reader should keep in mind that this book covers many areas of research that are rapidly changing. In some of these cases, the work is still very preliminary. I have attempted to present our best current understanding of what's happening, and I hope readers will pair enthusiasm for the progress scientists have made so far with a realization that there is still much more work to be done. I encourage the reader to check this book's website at www.systematicbook.com for updates on the topics discussed herein.

Contents

Preface: The Big Idea

Neurons are the cells that do most of the heavy lifting in your brain, but the most exciting thing a single neuron can do is fire an electrical pulse. Big deal. So can my toaster. But 86 billion neurons connected together in just the right way form your brain, which enables you to think, feel, imagine, and wonder. That's pretty impressive for a three-pound hunk of cells.

The difference between one neuron and your brain is more than a matter of scale. Those 86 billion neurons must be connected together, and then something fundamental changes. This book is about the magic that happens when many proteins, cells, or other biological pieces are connected into a system.

In biology, connected systems underlie many complex behaviors that have historically been hard to understand. Microbes in your gut, for instance, interact closely with your body, and some evidence suggests this system might affect your weight, alter your risk of developing cancer, or even influence your mood. Bacteria can use simple systems of proteins, DNA, and other biological material to anticipate sunrise, to decide what to eat, and even to predict what kind of food they are likely to encounter next. Systems in other living things affect aging, development, and overall health. And thinking in terms of systems can help explain interactions between organisms, such as the spread of plague through gerbil populations in the deserts of Kazakhstan.

Unfortunately for the impatient scientist, the systems found in biology are enormous and complex. Each person has about 40 trillion cells,[1] and each cell is made of billions of tiny components.[2] If you bought a gumball for every cell in your body, you could fill Fenway Park in Boston about 1,000 times over,[3] and those gumballs would cost you the entire GDP of Russia if you ordered them from Amazon.com at current bulk rates.[4] When this many parts are connected together in one big biological system, astoundingly complex behavior results. Even simple systems of just three or four components can produce sophisticated behaviors, and it is precisely these webs of interacting components that make life possible.

Systems biology—the study of connected groups of biological parts that all work together—is relatively new. For example, the Harvard Systems Biology Ph.D. Program matriculated its first class in 2005. There is even still some disagreement about what it means to be a systems biologist. Some think it's all about bringing the math back to biology research; others say it's about working with huge amounts of data. But at its core, systems biology is simply the recognition that life is complex because it's connected. Systems biologists want to understand how systems make life—and all of its weirdness—possible.

This is a book about how understanding natural systems is helping us unravel some of the biggest mysteries in science. We will explore biological systems that range in size from microscopic proteins to entire ecosystems. Despite superficial differences, these systems can be studied with similar approaches, and all of them have implications for our understanding of life or our ability to treat diseases.

PART I
THE BASICS

CHAPTER ONE

Seeing the Systems in Biology

Technological Advances Are Letting Scientists Understand Living Things in a New Way

Biologists were studying systems long before the term "systems biology" existed. Indeed, all living things are made of systems. The human body, for example, is a system made of organs—such as the heart, skin, and brain—that work together to make life possible. In turn, each of these organs is itself a system made of specialized cells that coordinate to pump blood, heal wounds, or sense pain. And each of those cells is a system made of proteins, DNA, and other microscopic biological parts.

But while scientists have always studied systems, their capacity for fully understanding them has been limited historically. A complete accounting of how a system works requires not only knowledge of each of its parts, but also an understanding of how those parts interact to make the system work. In many cases, we haven't even had the technology to measure all of the system's components; it was like doing a jigsaw puzzle with half of the pieces missing.

To make progress, researchers started by breaking systems down into little pieces and studying each part individually—doing biology from the bottom up. Some of today's scientists devoted their entire doctoral dissertations to trying to understand a single gene or protein, and for good reason: deconstructing systems into their constituent components has taught us much of what we know about biology today. But we now also have the capability to start putting those pieces back together to understand how they make life work. That's systems biology.

To finally have the ability to grapple with systems in

earnest took great effort by many scientists, such as Dr. Eric Wieschaus.[1] As a young researcher at the European Molecular Biology Laboratory in Heidelberg in the early 1970s, Wieschaus was focused on understanding the biological machinery that governs fruit fly development. Fruit flies, like most complex life, start out as a fertilized egg, and Wieschaus wanted to know how the complex adult fly can come from such a simple beginning.

At the time, we didn't know which genes were involved in the system that governs fly development, so Wieschaus and his colleague Dr. Christiane Nüsslein-Volhard decided to find out. They mutated a bunch of flies using chemicals that damage DNA, hoping that damage would, by random chance, happen to disrupt a gene that was important for development. Wieschaus and Nüsslein-Volhard then painstakingly examined the progeny of these mutated flies, looking for those with developmental defects. They repeated this process over and over, hoping to mutate enough flies that they would be likely to disrupt most of the genes in the fly at some point. The researchers looked at thousands and thousands of mutant flies—about 27,000 genetic variants in total—and meticulously documented how each embryo's particular genetic mutation messed up its development. Wieschaus estimates that it took them six to eight months of long shifts looking at fly embryos under a microscope, but ultimately their patience was rewarded: the experiment revealed the vast majority of genes that control fly development. Their work won them the 1995 Nobel Prize in Physiology or Medicine along with another geneticist, Dr. Edward B. Lewis.

Wieschaus and Nüsslein-Volhard's experiment gave them a first draft of the list of parts that compose the fly development system. They also had some good luck: it turned out that many of the mutations they were making had easily interpretable consequences, so they could guess the function of some genes by seeing how a change in that gene affected development. For example, they found that mutations in one gene affected the development of a set of regular segments that form early in the fly's development. These mutated flies had only developed the odd-numbered segments and were missing the even-numbered segments. Based on these observations, Wieschaus and Nüsslein-Volhard could guess that this gene, now called *even-skipped*, helped create those even-numbered segments in normal flies. This type of information allowed scientists to start piecing together how genes might interact in order to produce the complex patterns of an adult fly.

With Wieschaus and Nüsslein-Volhard's foundational work and additional research by many other scientists, the challenge of understanding fly development shifted. We no longer had to ask which biological parts were involved in this system. Instead, we could ask: How do these parts work together to produce the patterns we see? And what can that answer tell us about how humans develop? These are the kinds of questions that interest systems biologists.

In other cases, scientists did systems biology by making detailed observations of a system's behavior and using brilliant mathematical intuition to deduce how the system works. For example, two decades before Wieschaus and Nüsslein-Volhard were poring over their flies, Alan Hodgkin and Andrew Huxley, biophysicists at the University of Cambridge, were carefully inserting silver wires into one of a squid's neurons to try to understand how it fires an electrical pulse. The neuron Hodgkin and Huxley used is unusually large—large enough to allow direct measurement of its electrical activity using tiny wires and a device called a voltage clamp that allowed the researchers to manipulate the flow of electricity in the cell.

At the time, scientists knew that the electrical activity produced by a neuron was caused by a flow of ions—electrically charged atoms, in this case mostly sodium and potassium—into and out of the cell. The exact details of the process and the cellular components that controlled this flow were still a mystery, however. Rather than trying to isolate all of the components of the system individually, Hodgkin and Huxley focused on gathering an enormous amount of data about how the system as a whole behaved. They measured the neuron's electrical activity in dozens of different conditions as they altered the voltage across the neuron's cell membrane using their electrodes. By thinking of the neuron as an electrical circuit and applying what they knew from physics, they were able to deduce how the neuron fires an electric pulse, known as an action potential. Hodgkin and Huxley described this process using mathematical equations—which they solved using a calculating machine that had to be cranked by hand—that predicted electrical spikes similar to those observed in real neurons. Their equations even foreshadowed some properties of then-hypothetical structures called ion channels—which we now know to be proteins that open and close to let ions pass through the cell membrane—long before these channels were probed by other means. Their mathematical description of this process, known as the Hodgkin-Huxley model, is still in use today, and they won the 1963 Nobel Prize in Physiology or Medicine.

Hodgkin and Huxley's work is an example of what systems biology often aspires to be—a concrete, mathematical understanding of a complicated biological process—but successes like theirs were relatively rare even as recently as the 1990s. Scientists were often stymied by systems that had more parts or were more complex than the firing of a neuron's action potential.

Things began to change around the year 2000 when a combination of exponentially increasing computing power and new experimental techniques made systems biology more practical. Faster computers allowed scientists to look for patterns in large data sets that were too big for humans to work with. At the same time, scientists and engineers were inventing new tools, such as DNA and RNA sequencing, that could produce data at that scale. Scientists have long been able to measure how strongly a single gene is "turned on," for example, but measuring how every gene is behaving in many different cells simultaneously was impossible until recently. These days, the technology to get large-scale information about every gene or every protein in cells is commonplace, and scientists have made similar breakthroughs in their abilities to measure other aspects of systems. We're increasingly limited not by our technical capabilities but by our ingenuity.

The advent of these new tools made modern systems biology possible, but it also changed the kinds of skills biologists need to succeed. Since math and computer power are so important to many modern laboratories, there has been a flood of talent from physics and mathematics.

David Botstein, then the director of the Lewis-Sigler Institute for Integrative Genomics at Princeton University, was one of the scientists who recognized early on that the same technologies that made systems biology possible were moving all biology in a more quantitative direction. Aware that college biology students were not getting the math, physics, and computer skills that they would need, Botstein created a new approach to teaching budding scientists in the early 2000s. His course, launched with physics professor William Bialek, is called Integrated Science. This program is an evolution of earlier courses at Stanford and the Massachusetts Institute of Technology, and it combines physics, biology, chemistry, and computer science into one intense course that teaches students how to think like a research scientist.

As a freshman at Princeton in 2008, I became one of those students. In an Integrated Science introductory session, I listened with rapt attention as Botstein articulated how scientists were using physics, math, chemistry, and computer science to solve the most interesting problems in biology. This interdisciplinary approach enabled researchers to study fascinating phenomena, and they used mathematical methods to help them understand what was really going on. For systems biologists, those tools were illuminating systems that had not previously been accessible.

CHAPTER TWO

Déjà Vu All Over Again

The Common Patterns and Principles of Natural Systems

When I lived in Germany in 2009 for a summer doing research at the European Molecular Biology Laboratory, I had a tiny problem: I didn't speak German. The science all happened in English, so that was easy. But if I wanted to make friends, navigate the transit system, or buy food? That was a bit harder. I'm ashamed to say I never really did get good at German, but the experience taught me a valuable lesson.

A few weeks into the summer, my coworkers invited me to see *Terminator Salvation*, the sequel starring Christian Bale. The film was in German, without subtitles. I was afraid I would be totally lost, but a Hollywood movie isn't too hard to understand on the macro-level. The good guys look like good guys. The bad guys look like evil robots.[1] Even though the details were different, I knew the formula from seeing similar movies.

Just like Hollywood movies, systems biology has a common language. Biologists study processes that seem very different on the surface: how neurons link up to form the brain, how simple molecules assemble to form a cell's "skeleton," how proteins interact to warn a cell that its environment is too salty, and how infectious bacteria coordinate to attack only when they have sufficient numbers to overwhelm their host. But even though these processes operate at wildly different scales and depend on unrelated parts, they often contain common patterns and operate according to similar principles.

In the early 2000s, Dr. Uri Alon's laboratory at the Weizmann Institute of Science in Israel was searching for these types of patterns in naturally occurring networks.[2] The researchers collected a half-dozen examples of biological systems—networks of neurons, proteins, and genes—but also some examples of nonbiological networks, such as web pages connected by links, transistors connected by wires in computer chips, and even kindergarteners connected by friendship. (Alon explained, "You go to a kindergarten and ask children: who are your friends? One of the tragedies of life is if X likes Y, Y doesn't always like X.") They then applied a simple mathematical analysis to search their collection of seemingly dissimilar networks for patterns.

The researchers first needed to write down these networks in a common language. Alon's group used a simple representation of their networks that consists of components and arrows. In the case of a network of genes where the activity of gene X turns on gene Y, they would draw an arrow from X to Y:

X → Y

Figure 1: X activates Y.

Similarly, for a network of neurons, an arrow from X to Y means that neuron X sends signals to neuron Y. And for kindergarteners, an arrow represents a one-way friendship: the fact that child X likes child Y. Each network was ultimately represented by a large web of components connected by arrows, something like a much larger version of this hypothetical network:

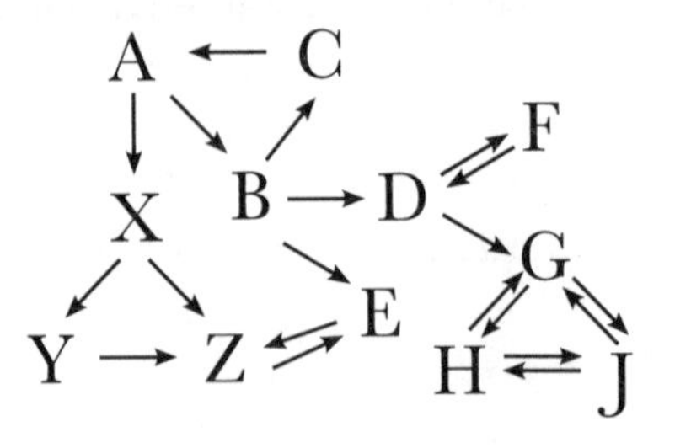

Figure 2: A hypothetical network.

Next, the researchers looked for patterns in these networks by examining groups of three or four components at a time. To start, they listed all of the possible ways that these sets of three or four components could be connected together. For example, a set of three components can be connected in any of 13 distinct ways. Here are three:

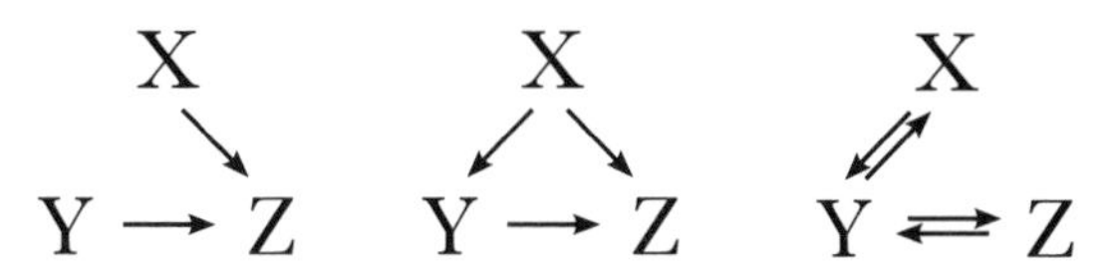

Figure 3: Different ways of connecting three components.

Alon's group then asked whether any of these ways of connecting the components shows up in real systems more often than one would expect by chance. Indeed, they found that certain sets of three or four components appeared far more often in the real networks than in randomly generated networks[3]—much in the same way that the three letters c, a, and t show up together in written English (cat) much more often than they would in a sequence of random letters. Alon and his colleagues call these repeated small patterns "network motifs."

In biological systems and some types of computer chips, for example, Alon and his collaborators often saw a network motif called the feed-forward loop. In this motif, X turns on both Y and Z, but Y also turns on Z. Schematically, it looks something like this:

Figure 4: A feed-forward loop.

Alon's team found that the feed-forward loop motif showed up frequently in networks of genes and proteins in bacteria and yeast, in networks of neurons in worms, and in certain kinds of electronic circuits. It seemed that the feed-forward loop often made an appearance in systems that process information—whether that was a bacterial cell decoding signals from its environment in order to move toward food, or a laptop interpreting a user's keystrokes. Indeed, Alon thinks that evolution and human engineers keep using these motifs because they work well: he says they may represent "the simplest, most efficient solutions to the shared problems that cells have." If that's true, the feed-forward loop would be the Michael Jordan of networks—the all-star who plays as much as possible because he gets the job done.

In the case of the feed-forward loop, one possible use for the motif is to help the network deal with noise—"noise" being, here, random fluctuations that could cause the network to make a mistake. If both X and Y must be turned on in order to activate Z, for example, the feed-forward loop could be used to turn on Z only after X has been active for some time—since the signal needs time to trickle down through Y—and to turn off Z quickly when X disappears. In a biological system, this behavior might be useful when dealing with noisy inputs, Alon suggests—if the organism might not want to respond with Z until it's sure the signal X is for real.

Long before Alon's group was analyzing kindergarten friendships and internet links, scientists had been prying apart biological systems to look for simple modules they could understand. In the 1960s, they began to find some simple circuits that seemed tractable, many of which were in bacteria. Even a single bacterial cell is exceedingly complicated, and we don't yet have the ability to understand the full dynamics of all of the stuff that goes on in the cell. But these early researchers made progress by finding smaller chunks of the bacterial system that were more manageable.

One early example of such a subsystem is the small circuit that allows the bacterium *Escherichia coli* to decide what to eat. In a pinch, they'll eat pretty much anything that has caloric value, but *E. coli* are actually picky eaters when food is plentiful—like teenagers who choose pepperoni pizza over a salad. Specifically, the bacterium sometimes must make a choice between eating one of two types of sugar that are commonly available to it.

The bacteria's first choice is glucose, the pepperoni pizza of sugars. Glucose is easy to break down and turn into energy, so it's ideal for a bacterium that wants, above all else, to get energy as quickly as possible: this helps it to grow quickly and outcompete other nearby microbes. But *E. coli* also has the ability to eat lactose, a kind of sugar found in milk. *E. coli* has to do a little extra work to break down lactose in order to use it as an energy source, and it makes a special enzyme[4]—a protein that speeds up a chemical reaction—called β-gal that helps break down lactose. It doesn't make sense for *E. coli* to be constantly producing this enzyme if there's no lactose around, or if the bacterium has plenty of glucose that it's using instead.

So what's a picky bacterium to do to make sure it only makes β-gal at the appropriate time? It uses a very simple regulatory system to test for the presence of lactose and glucose and to make a decision based on that information.

"Tests for glucose" or "makes a decision" are just anthropomorphic ways of conveying what is actually a pretty simple idea on the molecular level. *E. coli* senses lactose through a protein that normally grabs on to the DNA at a specific spot. This part of the DNA contains the gene that tells the cell how to make the lactose-digesting enzyme, β-gal. When this protein is attached to this bit of DNA, the cell can't make β-gal:

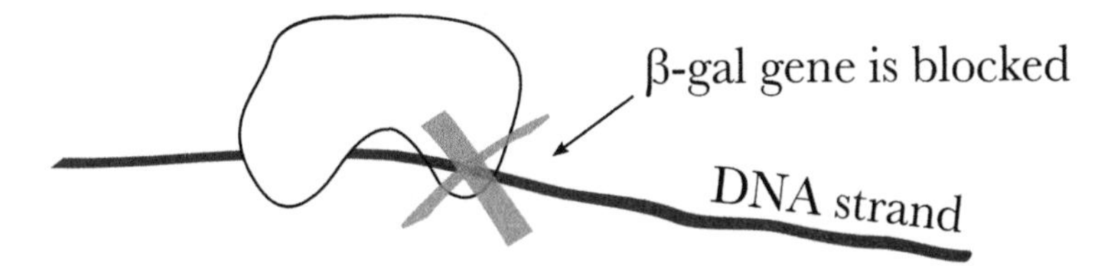

Figure 5: The blocking protein prevents the cell from making β-gal.

But when lactose is around, the protein grabs on to lactose[5] and changes its shape. That shape change prevents the protein from binding to the DNA, allowing β-gal production:

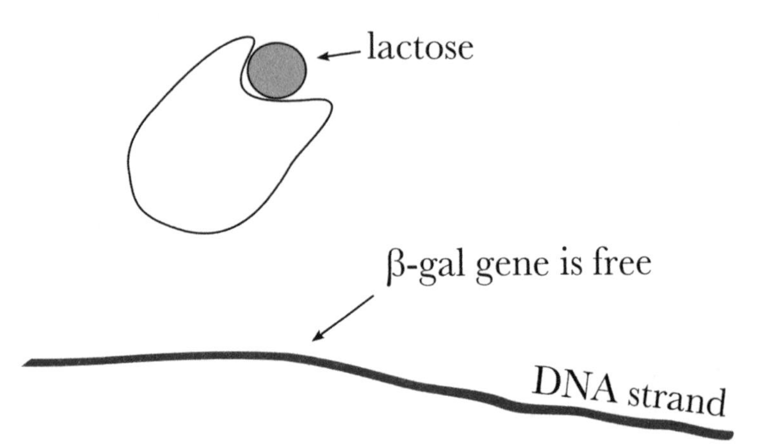

Figure 6: The cell can make β-gal freely when the blocking protein is bound to lactose instead of the DNA.

Since this protein can only grab on to lactose or the DNA at any one time, the cell can use this property to "sense" lactose.

The cell then also has to incorporate information about glucose into its decision. This is accomplished by a second branch of the system that prevents the lactose digestion from activating if there's glucose available. That second branch works in a similar way to the lactose-sensing branch. The cell senses whether glucose is available using a protein called CAP, which normally helps turn on β-gal production. But glucose's presence depletes a signaling molecule that CAP normally grabs on to in order to function properly. Without that signaling molecule, CAP doesn't work and the enzyme isn't made. While the first branch of the system said, "If there isn't any lactose, *don't* make β-gal," this branch of the system is saying, "If there isn't any glucose, *do* make β-gal."

Each of these branches responds individually to lactose and glucose, respectively, but taken together, these branches encode the logic we initially wanted: "If there is lactose and there is no glucose, make β-gal."[6] That's not a very complicated rule, but it does the trick for these bacteria. With this simple circuit, they can grow efficiently on any glucose that's available, then transition to utilizing lactose once the stores of glucose have been depleted. In fact, scientists can observe this shift happening. If you throw these bacteria into a tube with both glucose and lactose, they'll grow really quickly at first as they use the glucose. Then, as soon as the glucose gets used up, they pause for a moment, rev up those genes that let them use lactose as food,[7] and soon get back to growing quickly.

Even though this system performs a critical function for the bacterium, it is made of easy-to-understand parts. Any given component might grab on to lactose, or help promote gene expression (activation). But there isn't any one piece that says, "I'll check the glucose and lactose levels, decide which sugar we should be using, and then adjust the gene expression pattern accordingly." Instead, it is the system as a whole that controls which sugar these bacteria use.

While that decision seems pretty straightforward, even small systems can get complicated very quickly. Part of the reason for the glucose/lactose system's apparent simplicity is that I've described it as two linear pathways where A leads to B leads to C with no other inputs or outputs;[8] for example:

lactose → Unbinding of DNA-blocking protein → β-gal production

Figure 7: A linear system.

If that were the case for all biological systems, scientists would have a pretty easy time figuring them out. But when systems aren't just a simple straight line, everything gets much more interesting. Instead of A going to B going to C, imagine a feed-forward loop, or imagine that C can now loop back around to affect A:

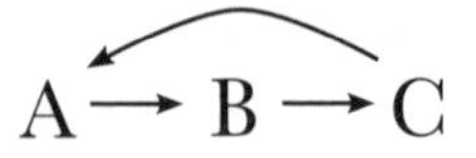

Figure 8: A feedback loop.

Or, A could affect both B and D, which both affect C:

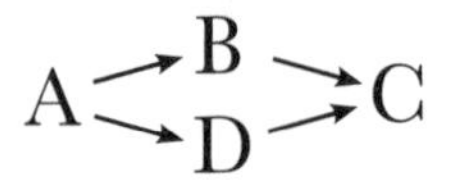

Figure 9: A branched system.

Depending on the details of the system, loops and forks like this can help regulate a system, create oscillations, prevent oscillations, make an on/off switch, or even produce literal chaos—a system where the behavior depends very finely on how the system starts out, commonly known as the butterfly effect. In any case, things get a lot more exciting.

Probably the simplest thing a loop can do is to allow a system to check on its progress. For example, some bacteria make proteins that work together to produce tryptophan, a basic cellular building block. (Tryptophan is the thing that supposedly makes you sleepy after eating too much turkey. This is a myth, by the way: it's overeating that makes you sleepy.) But once those tryptophan-making proteins have done their job and the cell has enough tryptophan, there's no need to make more tryptophan-producing proteins—and it would cost precious energy to do so. To avoid wasting energy, the system has evolved so that tryptophan circles back around and tells the cell to stop making the proteins that produced it. This feedback loop allows the cell to read how much tryptophan has been produced and adjust its behavior accordingly.

In addition to loops, systems can have another element that gives them extra capabilities: a delay. Indeed, the difference between protein A taking 1 second versus 10 seconds to turn on protein B can change the behavior of a system entirely. It's like jazz: sometimes the pauses are just as important as the music itself.

In fact, sometimes a very simple delay in a system can literally be the difference between life and death, as it is in the system that helps ensure cells make proteins accurately. The cells in the human body are constantly producing trillions and trillions of proteins. Those proteins require precise construction, but the cell somehow manages to make most of them correctly. Every protein is made up of little parts called amino acids, and these are connected in a long chain that then folds up like cellular origami to give the protein a specific shape that allows it to function. Those folded proteins then go on to do pretty much any cellular function that needs doing.

The information that tells your cells which amino acids to connect together and the order to connect them in comes from your DNA. DNA is made of long strings of four different components, or bases, that scientists abbreviate as G, C, A, and T. To make a particular protein, your cell makes a copy of the DNA instructions for that protein, then uses that copy as a template to tell the cell's protein-making factory which amino acid to add next with a simple code. For example, if the next three bases in the instructions are GCC, then an amino acid called alanine will be added to the protein. The molecule that carries that amino acid has a complementary sequence, CGG, that comes in and grabs on to the GCC sequence, and the carrier adds its amino acid to the end of the growing protein. This recognition happens because each base—each G, C, A, or T—has a few sticky bits that allow it to match up specifically with its partner; G sticks to C and A sticks to T.[9] Basically, making a protein correctly relies on successfully matching a sequence like GCC with its correct partner and then doing that same thing over and over again. That's a pretty fair description of what happens, but if that were all there was to it, we'd all be dead. Our cells would never be able to make a working protein because they would make too many mistakes.

When making a protein, the difference between the right sequence and the wrong sequence can be rather small. For example, the DNA sequence AAT tells the cell to use the amino acid asparagine, but the very similar sequence AAG codes for a different amino acid, lysine. Unfortunately, AAT looks a lot like AAG on a molecular level, so sometimes the wrong amino acid carrier grabs on to the template. Don't forget: the molecules that carry the amino acids can't think for themselves—they just stick to some sequences better than others. So if it were really just a matter of these amino acid carriers blindly sticking to the right three-letter code, sometimes the wrong amino acid would be added to the protein.

Actually, quite a few incorrect amino acids would be added. Scientists can use physics to estimate how frequently a mismatch would occur based on the energy difference between the code matching up with the right and wrong amino acids; this calculation indicates that we should expect to end up with the wrong amino acid 1 in 100 times, at the very best.[10] Since many proteins are hundreds or even thousands of amino acids long, almost every protein would have some mistake in it! And a mistake in the protein means that it might not work correctly, or it could even be harmful to the cell. Life as we know it would be pretty much impossible.

Indeed, we know from experiments that the actual error rates are much lower. We can estimate that the protein-making machinery in the cell only incorporates the wrong amino acid approximately one in every 10,000 times.[11] That's a much more manageable error rate, and this level of accuracy allows cells to make most proteins correctly.

This boost in accuracy is attributable to one final, critical step in the process: in order for the amino acid to be incorporated into the protein, it has to pause for a while. The amino-acid-carrier molecule will come along and bind to the sequence, and then it will sit around for a little bit before the final bond can be made to add the amino acid to the end of the growing protein chain. This "finalization" process takes a relatively long time, and it's that delay that improves the accuracy of the protein-making process considerably.

Basically, the delay gives an incorrect match more time to fall apart. It's like a bouncer checking IDs at the entrance to a bar. If you barely give him any time to check before he has to decide whether or not to let someone in, college kids with fake IDs are inevitably going to slip into the bar: the fakes look pretty close to the real thing. But give him time to examine the license really closely, and he'll catch most of the fakes. The longer he has to think, the more likely he'll be to reject a fake—so if you let him examine the license for five minutes, most of the people who have survived his scrutiny for that long have real IDs.

At first, this doesn't seem like a very good system; it's pretty inefficient. It takes a relatively long time to add another piece to the protein chain, and many of the potential matches, even the correct ones, fall off while waiting to be finalized. But while a lot of correct matches will fall off, even more bad matches will fall off. Thus, the cell doesn't make proteins quite as fast as it might otherwise be able to, but it gains some precious accuracy. When it finally does commit to adding an amino acid, the error rate is quite low—low enough to correctly make proteins and ultimately to allow for complex life as we know it.

In other contexts, delays can also cause systems to overreact to deviations or to oscillate, or they can be used to avoid responding to a signal until it's been sustained for some time. And to some extent, every connection in a real biological system comes with some delay, and that delay adds another wrinkle that makes systems a lot more complex.

Circadian Rhythm

In the competitive world of biology, timing is everything. Knowing when to hunt and when to sleep is critical to an organism's survival, and maintaining an internal clock helps it keep time with the natural daily cycle of light and dark. On a longer time scale, birds must mark the months to know when to migrate to escape winter. Butterflies seem to use molecular clocks to time their seasonal migrations, and they also appear to have clocks in their antennae that they use to help navigate to their migratory destinations.[12]

Biological clocks are also deeply embedded in the inner workings of our bodies. A daily cycle, called a circadian rhythm, governs our times of sleep and wakefulness. Messing with the day-night cycle or the genes that run the clock in mice causes a whole host of problems, including obesity.[13]

Biological clocks can be exceedingly simple. Some biological clocks simply involve three proteins interacting in a small system. In fact, the core of one such system from cyanobacteria is so simple that we can isolate the three proteins that make up the clock and put them in a test tube, and they will merrily go on keeping time for days.[14] This circadian clock centers on a protein called KaiC. The protein is modified by two other proteins, KaiA and KaiB, that add a small chemical marker onto KaiC or remove that marker. In the test tube as in the cyanobacteria, the percentage of the KaiC protein that is modified oscillates, peaking approximately every 24 hours. The cell watches how much KaiC is marked to know, for example, when the sun is about to rise: this allows the bacterium time to prepare to capture the energy it needs to grow at its maximum potential.

A clock is nothing more than an event that occurs periodically at reliable times, and this cycle of protein modification does just that. In fact, this regular oscillation is all we really need to make a clock and keep track of time. Perhaps the most familiar clock is the sun rising and setting, which ticks every twenty-four hours. Old-fashioned grandfather clocks used weights that would cycle back and forth at a constant rate. As for the Dora the Explorer Timex you might get in your Happy Meal? It probably counts the number of vibrations of a miniature quartz crystal, which resonates when current is passed through it.

There's one final common systems principle to consider: powerful behavior can result when parts work together, a phenomenon known as cooperativity. Cooperativity means that when one part of a system does something—binds a molecule, changes shape—it makes the other parts more likely to follow suit. Cooperativity happens in all sorts of systems, even in nonbiological networks: if all of your friends create an account on a social network, for example, you might be more likely to create an account, too. Any process where A happening makes B more likely to happen, and vice versa, could be thought of loosely as cooperativity.

A classic example of this power is the mechanism by which oxygen gets around the body. In humans, the oxygen we take in through our lungs binds to a protein called hemoglobin in our blood, which carries the oxygen to the tissues where it's needed. But hemoglobin has a tough job because in some ways it has to be two different things at once.

When hemoglobin leaves the lungs it is filled up with as much oxygen as it can carry and takes the oxygen to tissues that need it, such as an exercising muscle. It then dumps the oxygen into that tissue, and returns to the lungs to fill up again. One might think that, in order to load up on as much oxygen as possible in the lungs, hemoglobin should bind oxygen very tightly. But if that were the case, then it would be hard for the hemoglobin to give up the oxygen when it gets to the muscles. On the other hand, if hemoglobin bound oxygen weakly, it might not pick up enough oxygen in the lungs, or it could lose the oxygen before getting to the tissues that need oxygen the most. Ideally, hemoglobin would be able to hold on to oxygen tightly in the lungs and give it up easily when it is near a cell that needs oxygen.

The real problem is that hemoglobin is just a dumb molecule. Hemoglobin doesn't "know" a lung from a muscle, or a Red Cross blood bag. All hemoglobin knows is how much oxygen is available in its immediate environment. This suggests a possible strategy: hemoglobin should bind oxygen tightly when there's a lot of oxygen around, such as in the lungs, but it should give up all of its cargo when the concentration of oxygen gets low. Using this strategy, hemoglobin could deliver oxygen efficiently.

To implement its strategy, hemoglobin employs four pieces, called subunits, all of which communicate with one another. Each subunit holds on to one oxygen molecule. When one piece grabs onto an oxygen molecule, it makes the other pieces of that hemoglobin bind more tightly to oxygen than they would have otherwise. Basically, this cooperativity makes all of the oxygen binding sites want to be in the same state—either bound to oxygen or not. So if one subunit of the hemoglobin complex binds an oxygen molecule in the lungs, all of the others will want to as well. And if one subunit gives up its oxygen to a muscle cell, the others will likely follow suit.

This cooperative behavior means that hemoglobin does exactly the right thing: it tends to fill up on oxygen in environments with a lot of oxygen and release it more easily in low-oxygen environments. This dramatically improves the efficiency of oxygen transport, and it probably helped our ancestors do oxygen-intensive activities, like run after a gazelle, or run away from the angry lion that was also chasing that gazelle.

But not all animals transport oxygen this way. For example, some arthropods and mollusks use a molecule called hemocyanin, which in some species doesn't have the same cooperative oxygen binding behavior that human hemoglobin does. This makes their oxygen carrying capacity significantly less efficient than ours[15]—because our hemoglobin knows how to cooperate.

CHAPTER THREE

America's Next Top Mathematical Model

Understanding Complex Systems Sometimes Requires Math

For some simple systems, we can get a pretty good idea of what's going on just by using our intuition. In the previous chapter, for example, the system that allowed bacteria to choose between glucose and lactose was easy to describe. And after you learn how it works, you can make some simple predictions: "If a cell sees lactose but not glucose, it turns on the lactose-digestion enzyme."

But the written description belies some of this system's complex behavior. First, history matters: moving the cell from an environment with no lactose to one with a low concentration of lactose may not be enough to turn on the enzyme, β-gal, but the cell will produce the enzyme at that same low concentration if it is first treated with a high concentration of lactose:

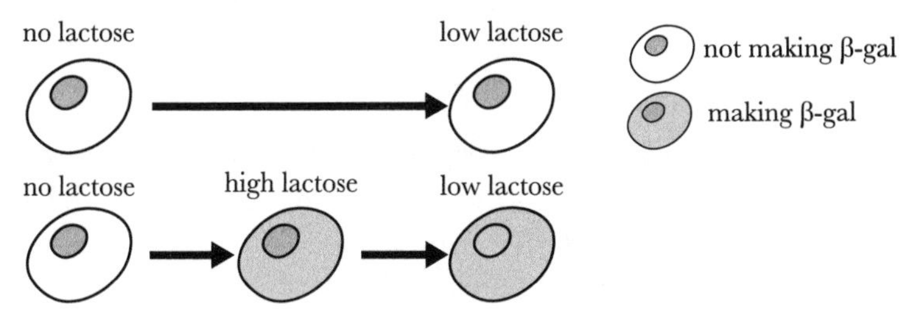

Figure 10: Past lactose availability can affect current β-gal production.

Second, glucose normally prevents the cell from making the enzyme even if lactose is around. But if the lactose digestion is activated first, a later influx of glucose may not shut it off:

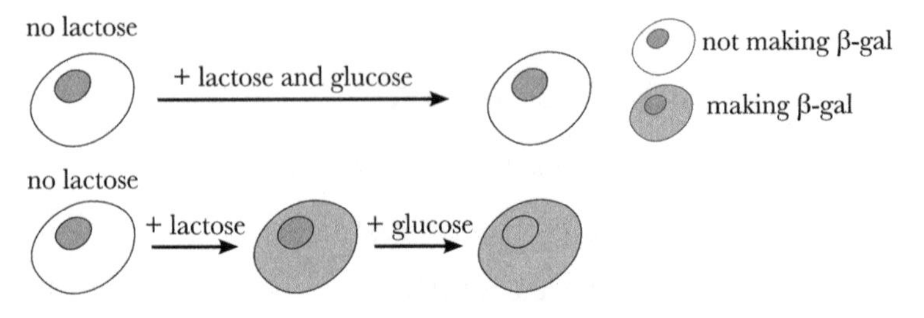

Figure 11: The relative timing of glucose and lactose signals can affect β-gal production.

And third, this system can exhibit a property known as bistability: either it's on or it's off. If there is a small amount of lactose signal, one might imagine that all of the cells respond weakly. In actuality, only some cells respond, but they respond fully:

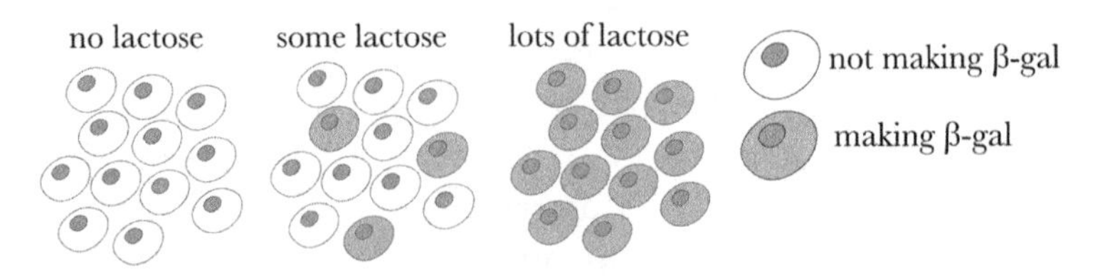

Figure 12: A group of cells can respond heterogeneously to the same amount of lactose.

In short, there's a lot more to even this simple system than meets the eye.

Since intuition fails us rather quickly as systems become more complex, we need a way of thinking about them that's a bit more rigorous. For systems biologists, this often means building a mathematical model.

A model in systems biology is just a mathematical description of how a system works. We can then use the model as a tool to make predictions and refine our understanding. A model might be something like this: "The cell makes protein A and protein B at a constant rate of 5 proteins per minute.

Both A and B proteins degrade at a constant rate and last for 10 minutes on average." Basically, a model is really just a set of rules. Using either math or computer simulations, we can use those rules to figure out how that model should act: "If this were how the system worked, this is how we'd expect it to behave."

Mathematical modeling forces researchers to write down how they think the system functions. They can then test this understanding by using the model to make predictions and compare those predictions to the actual behavior. And when the model is wrong, they refine it.

Every model is a simplification. Models of how predators affect prey populations ignore the details of how the predator metabolizes food into energy to allow it to give chase. Models of how signals are carried through the cell by proteins ignore the tiny atoms that those proteins are made of. Consequently, we always have to make choices about what to model and how to model it, based on which parts of the system we believe to be important.

When models become too complicated, they cease to be useful. Jeremy Gunawardena, an associate professor of systems biology at Harvard Medical School, told me about a time he and his collaborators were working with a system of about a dozen components. He wrote up all of the rules they thought might be important and then tried to make a model with a technique that should produce an equation that describes how the system works. With the help of a computer program, he found such an equation, but it was six pages long.[1] That's pretty hard to work with, and it's not very useful.

In the best case, models are simple, specific, predictive, and illustrative. The model of a neuron developed by Hodgkin and Huxley, mentioned in Chapter 1, is a prime example. By applying simple equations that would be familiar to any college physics student, Hodgkin and Huxley explained the behavior of a seemingly complex system, the squid neuron. Their equations made specific predictions that they could back up with experiments. And the model still shapes how scientists think about neurons today: more than 60 years later, scientists still use a modified version of their model, and it's a foundational framework for any neuroscientist.

Indeed, with very simple extensions, this model can be used to describe small circuits of neurons. For example, Brandeis University professor of biology Eve Marder studies a circuit that drives the rhythmic muscle contractions crabs and lobsters use to break down their food. Here's an example of the output from such a model for three neurons (called PM, LP, and PY) that fire repeatedly in sequence. The three traces represent the electrical potential of each of the three neurons, and you can see the sharp spikes that indicate firing events clustered into clumps:

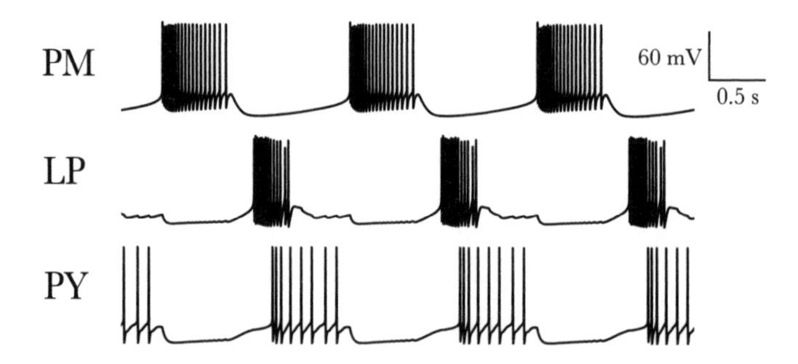

Figure 13: Simulated electrical activity of three neurons in a rhythmic circuit.

Marder's lab uses these types of models to investigate how different two "normal" brains might be from each other on the cellular level. (This will be covered in Chapter 10.)

In other domains, models have served to sharpen our questions. For example, scientists have studied the cabling that allows a dividing cell to separate its DNA into two halves since at least the 1950s.[2] These cables, known as the "spindle," move half of the DNA into one daughter cell and half into the other. The spindle in a cell that is about to divide looks something like this:

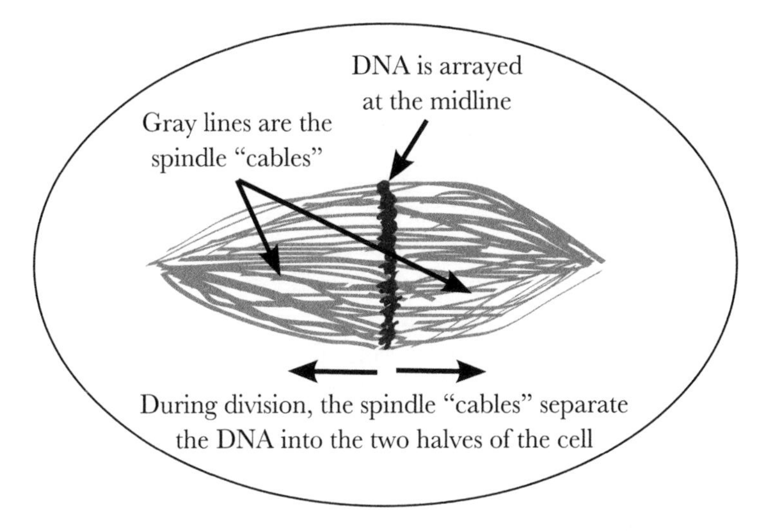

Figure 14: A ready-to-divide spindle.

Each half of the spindle then contracts, dividing the DNA and making sure that the two daughter cells will both have a full copy of the genetic information:

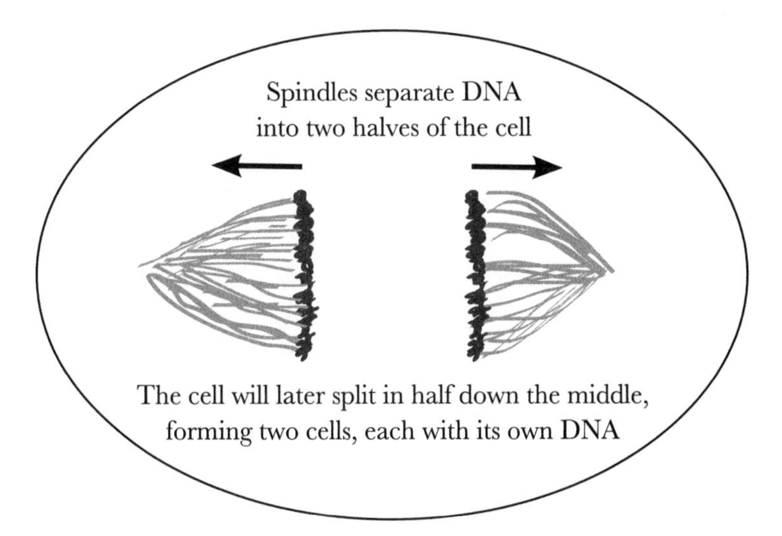

Figure 15: The spindle divides the DNA into the two daughter cells.

Since we know the parts that make up the spindle and we can describe how the spindle behaves—it separates the DNA into the two daughter cells—it might be tempting to say that we understand the spindle. But that simplistic understanding leaves a lot of unanswered questions that become apparent when one tries to build a mathematical model of the spindle. How exactly does the spindle build itself? What sets the size of the spindle? Which interactions between the individual cables are important, and which are not?

In 2014, Daniel Needleman, an associate professor of applied physics and of molecular and cellular biology at Harvard University, and Jan Brugués, a postdoctoral fellow in Needleman's lab, developed a mathematical model of the spindle that answers questions like these.[3] The researchers made careful measurements of the density of the spindle and how the fibers are oriented, among other things, and they used physics inspired by that which has previously been used to describe the material that drives the LCD (liquid crystal display) screens found on some televisions and phones. Using their model and experimental measurements, they were able to show that simple rules—such as the tendency of two neighboring fibers to align with each other—are enough to explain the shape and behavior of the spindle. This model provides a much more complete and precise understanding of how the spindle works than was previously available.

On a much larger scale, mathematical models can also provide a framework for thinking about ecosystems and the natural world, such as equations that describe the relationship between predators and their prey.

And of course, models can describe behaviors that our intuition fails to illuminate. Consider a phenomenon named after one of the world's greatest codebreakers: Alan Turing. Turing is most famous for his work in the fields of computer science and encryption. He was a huge force behind cracking the German Enigma code during World War Two.

But Turing was also one of the first to recognize a counterintuitive phenomenon that can produce complex patterns from simple systems. The basic idea involves diffusion, the process by which two fluids spread out and mix together; you can see diffusion at work when you drop food coloring into water or cream into coffee.[4] Normally, we think of diffusion as a calming force: it turns a drop of cream in coffee into a well-mixed drink. But this mixing process can also sometimes make an otherwise stable system exhibit dynamic behavior. This destabilization, known as the Turing instability, can produce complex patterns. And in some cases, it's possible that living things use these patterns to define structures, such as the shape of a person's fingers.

Dr. Erwin Frey, the chair of statistical and biological physics at Ludwig-Maximilians-Universität in Munich, uses a simple analogy to explain the Turing instability. It's a tale of sheep and wolves. (Even though real sheep and wolves don't behave this way, the names will make for an easier story.) Imagine some sheep and wolves start out grouped closely together in the center of a large field:

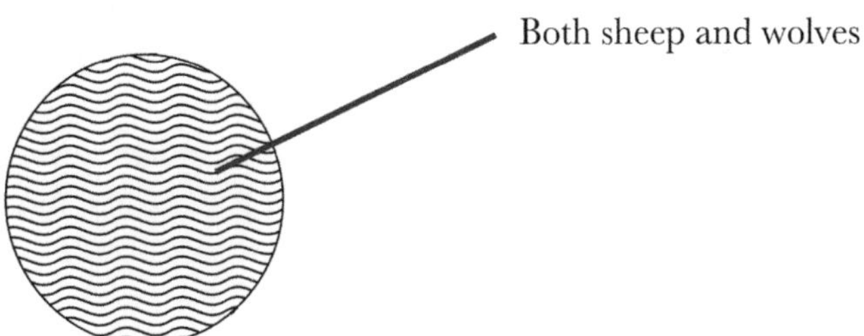

Figure 16: The sheep and wolves start out clumped together.

Both species start to wander around randomly, which causes both groups to spread out. Wolves spread out quickly, while sheep spread out more slowly:

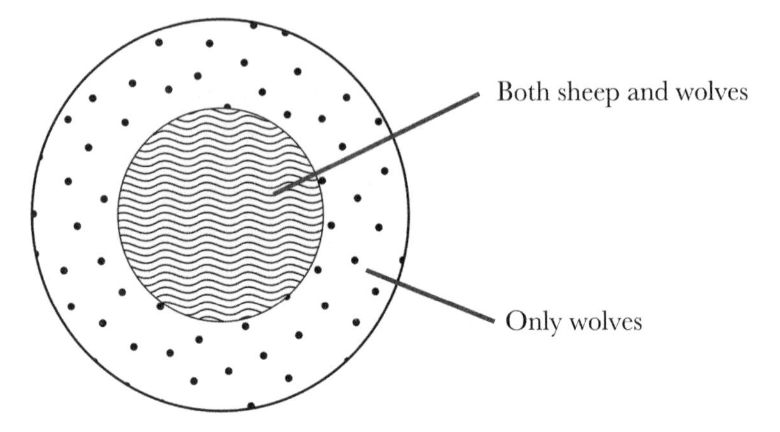

Figure 17: The wolves spread out more quickly than the sheep do.

While this spreading is happening, both species are breeding, and sheep breed faster than wolves. Now you have a lot of sheep in the center of the field, since they breed quickly and don't spread out much, and there are a smaller number of wolves dispersed in a wider circle that surrounds the sheep. Next, the wolves eat sheep. This doesn't have a big impact on the sheep population in the center of the field, since they breed quickly and there are already so many of them, but they are eaten and overwhelmed by wolves if they try to spread out from their clump in the center of the field. This interaction creates a stable pattern where there is a single region of sheep that are bordered and kept in place by the wolves that surround them. As a result, there is one "spot" of sheep.

The situation would be a bit less intuitive if all of the sheep and wolves had not started in the center of the field, but the same logic leads to complex patterns. The results are easiest to see using a computer simulation. In the following images, black regions are areas with a lot of sheep while white areas have few sheep. The simulation starts off in a noisy initial configuration with both species roughly evenly distributed throughout the available space:

Figure 18: At the start of the simulation, the sheep are spread roughly evenly across the field.

After a short amount of time has passed, there are already hints of a pattern, as the sheep clump up into stable spots:

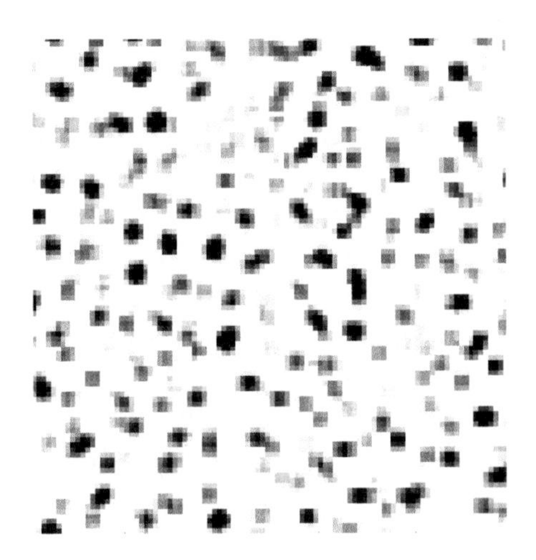

Figure 19: Spots begin to appear.

And after a bit more time, the system exhibits a stable, sharp pattern of sheep spots:

Figure 20: A clear pattern emerges.

Any types of components—whether they are species, proteins, or chemicals—can display this kind of behavior. The patterns shown above depend just on how the components behave, not on what they are: if they behave like "sheep" and "wolves," similar behaviors will emerge. Broadly speaking, the sheep-wolf field is a type of a general class of systems called reaction-diffusion systems. In these systems, simple interactions between components combined with the ability for each of the components to "spread out" causes striking patterns to emerge. Depending on the details of the system, such as how fast the animals spread out and how strongly they interact, these patterns can take the form of spots, stripes, swirls, or even waves that travel back and forth over time.

Scientists have hypothesized that all sorts of biological patterns may have something to do with the Turing instability, but that has been hard to show in practice. Frey says that's because the Turing instability, strictly defined, requires a few things that rarely happen in real biology. One of the two species has to spread out much faster than the other, for example, and each of the components must "breed"—that is, stimulate the production of more of that same component—on the right time scales relative to the pattern formation.

Some evidence suggests that a three-molecule reaction-diffusion system may play an important role in limb development, or more specifically, how our fingers and toes came to be.[5] Basically, there is a point during development when a fetus has to go from having just a stub at the end of a proto-arm to having fingers. There is a molecule that indicates "this should be a finger," so where there was a lot of that molecule, a finger developed, and where the molecule was absent, a gap appeared. By using a marker that lights up under a microscope, scientists can look at a developing limb and see stripes of the finger-making molecule, and some scientists think that a reaction-diffusion system forms those stripes. And when this goes wrong, one can end up with too many or too few fingers or toes.

Another simple model with complex behavior comes to biology from physics: the Ising model. A Ph.D. student in Germany in the 1920s, Ernst Ising, and his adviser, Wilhelm Lenz, used this model to describe ferromagnetism, the type of magnetism seen in refrigerator magnets. On a microscopic level, a refrigerator magnet looks something like a checkerboard with mini-magnets on each square. Each little magnet can be in one of two states—up or down—so we can represent each little magnet as an arrow (see Figure 21).

A material becomes noticeably magnetic when all of these little magnets line up with one another and point in the same direction: all of the little magnetic fields combine to form one big one. In contrast, if the arrows are pointing up or down at random, the small magnetic field of each mini-magnet is canceled out by its neighbors.

These little magnets interact with one another, and each wants to align itself with its neighbors. Each tiny magnet

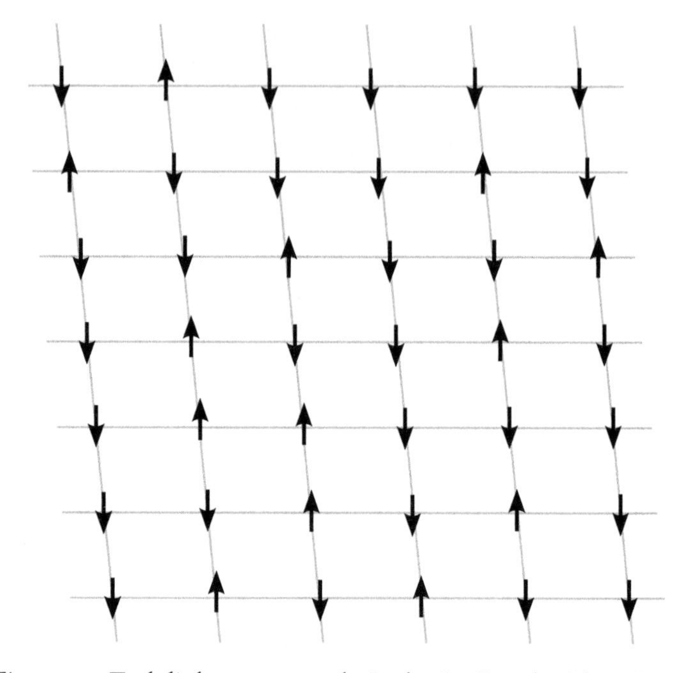

Figure 21: Each little magnet can be in the "up" or the "down" state.

is also jiggling around with thermal energy, however: temperature is just a measure of how much each particle in a substance is vibrating around on a microscopic scale. At high temperatures, the mini-magnets jiggle around a lot, and the interactions between them are dwarfed by the thermal energy. Each small magnet is jiggling around so much that it points up or down at random, and there is no visible structure and no overall magnetic field. In Figure 22, black pixels represent magnets that are in an "up" state, and white pixels represent a "down" state.

But at lower temperatures, there is less thermal energy, so each small magnet isn't jiggling around as much. Since each mini-magnet still wants to align itself with its neighbors, this tendency starts to overpower the thermal energy. This causes the magnets to orient in local patches (see Figure 23).

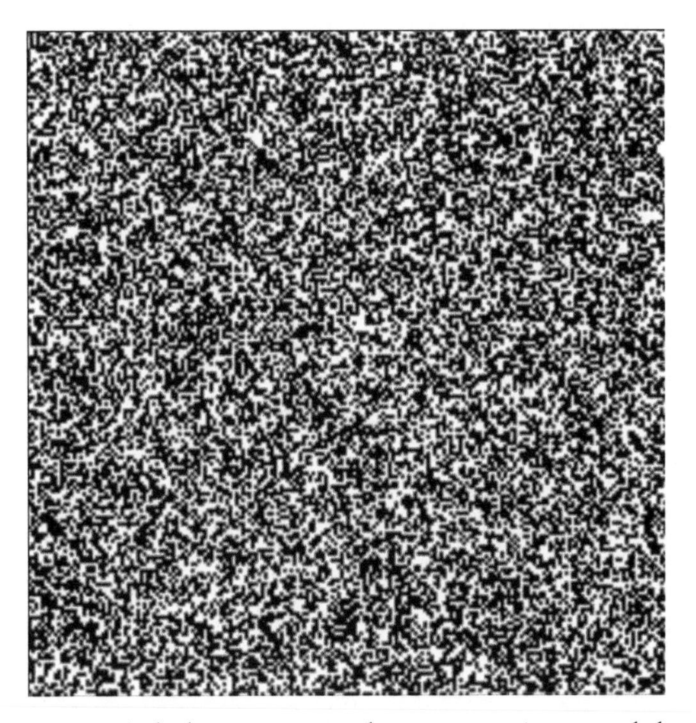

Figure 22: At high temperatures, the magnets point up and down randomly.

The appearance of these patches is known as a "phase transition," and it's similar to the phase transition that happens when water freezes: just like each mini-magnet aligns with its neighbors, water molecules lock together in a rigid structure when they form ice. The Ising model is one of the simplest models that displays such a transition.

In biology, the Ising model and other similar frameworks have been used to describe many different systems. When it's used to model networks of neurons,[6] the "up" or "down" state of each grid square indicates whether a neuron is "on"—firing electrical pulses—or "off." Ising-like models have been used to study how proteins in the cell membrane communicate with one another, and they have been used to describe small molecules binding to long chains, such as the proteins

Figure 23: At low temperatures, the magnets align with their neighbors.

that form a cell's structural skeleton.[7] One of my friends from college has even used a similar framework to study the U.S. Supreme Court.[8] In his model, each magnet represents one of the justices, and the "up" or "down" state of the magnet corresponds to the way that justice votes on a case.

Biologists have used modeling to help them understand small systems with great success. But many of the things that are most interesting to study, such as a whole cell, have thousands of parts that would probably be important to include in a model. The scientists who build models of large systems hope that they will help us identify gaps in our knowledge, predict the effects of disruptions on the behavior of the system, and provide a framework for understanding new data and engineering novel organisms.[9] These goals are appealing, but big models also pose big challenges.

Take the model that some scientists built for the tiny organism *Mycoplasma genitalium*, for example. *Mycoplasma* is a comparatively simple organism. It's composed of a single cell with only 525 genes. It doesn't have very much DNA, and each *Mycoplasma* cell is something like 10% of the size of a regular bacterium.[10] But compared to the scale of typical models, it's huge.

The researchers, led by Markus Covert, associate professor of bioengineering and of chemical and systems biology at Stanford University, built a mathematical model of how the whole cell works. Essentially, they took the processes that they thought were most important in the cell and broke them down into 28 different, smaller models.[11] These little models were simulated independently for short periods. Then those smaller models were all allowed to influence one another, and the process was repeated many times. Each sub-model basically describes how much of each component in that subsystem is present at any given time. Together, these subsystem models represent everything that's going on in the cell.

When simulated, the model does things that resemble what a real cell does. For example, the virtual cell divides to make two new cells at approximately the right rate. The overall levels of cellular parts are similar to the levels scientists see in real cells. And when the researchers deleted a gene from the model to see if the virtual cell would still grow, the results matched the observed viability of real cells most of the time. The model misses many of the finer details, though.

The true test of the utility of this model and others like it will be their ability to make specific new predictions that can then be validated experimentally. Covert's team has made some initial progress in this regard by investigating three cases where their model incorrectly predicted that eliminating a specific protein would be fatal to the cell. By looking at other proteins in their model that might be able to compensate for this loss, the researchers were able to predict that their parameter estimates for these compensating enzymes were incorrect. They then predicted what the correct values should be, and those predictions matched the results they got from measuring those values in real cells pretty well.[12]

But there's a general challenge that scientists must deal with as models get bigger and bigger—a challenge that's common to all models. Every time we add something new to the model, we introduce more flexibility. In modeling, flexibility increases the range of behaviors the model can predict, which makes it more likely that you can produce the right behavior with the wrong model. The physicist John von Neumann used to joke that with four parameters he could make a model produce the shape of an elephant, and with five he could make the trunk wiggle;[13] with enough free parameters and enough fiddling, you can make a model do anything you want—which makes it impossible to tell if your model is at all correct!

But for many biological systems, dealing with a lot of parameters may be unavoidable. The *Mycoplasma* model, for example, has about 1,900 parameters that describe everything from the speed of chemical reactions to the total amount of DNA in the cell. Detailed models of more complicated cells would require many, many thousands of parameters.

One strategy for dealing with a model's flexibility is to try to measure the real-life values of the parameters in the model. The creators of the *Mycoplasma* model chose the values of their parameters by finding published experiments that had measured many of the parameters they needed or by measuring those values themselves. This practice constrains the model considerably and makes it less flexible—and therefore more able to make specific predictions.

But these measured values can be subject to experimental error. What if one of the chemical reactions is accidentally set to be twice as fast as it should be—or several hundred times as fast? It's impossible to get all of the parameters right in a model that large, so it's reasonable to ask how correct the parameters need to be in order to get a useful model.

It's not even clear that there's a single "right" value for all parameters. Many cellular parameters vary with temperature, pH, and cellular context, so the correct values may depend on the current conditions. Heterogeneity in the population could also be a problem; imagine a cell with one parameter where the cell can stay alive only in one of two conditions: either parameter A is very high or it is very low. In a population of many cells, some may have high A and some may have low A. Since many of the methods scientists might choose to measure A report the average value of that parameter across a whole population of cells, scientists might accidentally choose a medium value for A when modeling those cells, even though no single cell actually has that value—and having a medium value for A might even kill the cell.

Not all parameters are equally important to the ultimate behavior of the system, however. If we increase parameter X but compensate by decreasing parameters Y and Z, we may have a relatively small effect on the system's behavior: this is called a "soft" combination of parameters because the system is broadly insensitive to changes of this type. On the other hand, a model is very sensitive to changes of "stiff" combinations of parameters.

In many large, real-world systems, a fortuitous simplification tends to occur: only a handful of parameter combinations are stiff, the system being broadly insensitive to most ways the parameters could change. This pattern holds for systems as diverse as systems of proteins, a variation of the Ising model, and networks of neurons.[14] This simplification allows scientists to make predictions about how a model should behave, even when they don't know all of the parameters.[15] These predictions focus on changes in behavior: "If the model is right, the system should be sensitive to changes in these parameters and insensitive to these other changes."

Besides increasing our ability to study large systems, thinking about stiff parameter sets allows us to get at the crux of a system's overall behavior: Which parts and which interactions are actually important? It also gives us some reason to be optimistic about the future utility of relatively big models: we might not have to get everything right—just the things that matter.

CHAPTER FOUR

Ignoring the Devil in the Details

Robustness, Prediction, Noise, and the General Properties of Systems

Studying a system can sometimes feel like buying a remote-controlled toy helicopter for a child. The buyer must evaluate a complicated toy not by understanding all of its parts, but based on a few large-scale properties: How long can it fly on a charge? How fast can it go? Will it break down often and cause temper tantrums? These are the things we really care about, not whether transistor #3546 is connected to #2143 or to #7824.

Scientists, too, are often concerned with the big-picture results. Instead of worrying about every single detail, we can sometimes take a step back and ask about the important large-scale properties of a system: Does the system respond well to disruptions? How much random variation can it tolerate? Do all of the parts matter equally to the system's overall behavior? Does the system break if one of those parts is slightly modified?

Consider a system's ability to keep functioning despite disruptions to its inner machinery. Many of the challenges that plague large mathematical models of biological systems are also problems for the systems themselves. A modeler might have trouble getting thousands of parameters exactly right, and so does a cell. Biological systems are messy, and life wouldn't be possible if everything had to be exactly right every time. But ultimately, a cell doesn't care about how quickly its demethoxyubiquinone hydroxylase chews through 5-demethoxyubiquinone; it cares about staying alive and reproducing.

A robust system—one whose overall behavior is insensitive to the details of its parts—has a clear advantage: it can tolerate unexpected imperfections in the system without falling apart entirely. If the behavior of the network depended very precisely on how fast a given reaction happens, for example, cells would break quite often; even a simple change in temperature alters the rates of chemical reactions because increased thermal energy causes everything to speed up.

A system might also be robust to changes in some parts but not in others; it's possible that it doesn't matter if the system converts one molecule to another at a rate of 50 molecules every second or 50,000 molecules per second, while another part must be carefully tuned.

One classic example of a robust system is the circuitry that allows bacteria to find their way around. These microbes navigate by a primitive sense of smell: their food gives off certain chemicals, and they then follow those chemicals to the food. Unfortunately, bacteria are so small that they can't tell which direction the smell is coming from just by standing still; the smell is basically the same all around something the size of a bacterium.

Instead, the bacterium has to explore. It starts out by moving randomly in one direction. If things are getting better—that is, if the smell of food is getting stronger—it keeps moving in that direction for a while. But if things are getting worse, it stops, "tumbles" around a bit to choose a random direction, then tries going that way. On average, this lets the bacterium "run and tumble" its way toward the food. All the bacterium has to do is "tumble" less often when the smell is getting stronger.

Scientists have studied many aspects of the "run and tumble" behavior, including its ability to adapt. In this context, adaptation means that the bacterium becomes accustomed to different smell strengths. Imagine that the bacterium starts out far away from the food, swims in a random direction, and finds a higher concentration of the smell molecules. Initially, it will want to reduce how often it tumbles: the smell of food is getting stronger, so it should keep going in this direction for a while. But eventually, if the concentration of smell stays the same, it will start tumbling more frequently again to see if it can get even closer to the food. The bacterium "adapts" to an increase in the attractive smell; this allows it to always move in the right direction regardless of whether it's in a region with a lot or just a little of the smell. Put another way, if we were to dump a bunch of food-scented chemical onto a bacterium, it would initially tumble very infrequently, then eventually return to whatever baseline tumbling frequency it had.

This property of adaptation turns out to be robust. When scientists manipulated the individual components that make up the system by forcing the cell to make more or less of that component, the system was still able to adapt to new smell concentrations;[1] it always returned to its baseline tumbling rate.

And while adaptation is robust, other behaviors of the bacterial find-the-food system—such as the exact value of the baseline tumbling rate or the time it takes the system to adapt—*do* vary widely depending on how the individual components are tuned. In this case, it seems that the thing that is most critical to the function of the system, the adaptation, is robust, and the other properties of the system are able to vary.

While robustness is all about making sure the system functions when things inside it change, not all challenges come from within. Real-life biological systems face challenges from their environments all the time. For situations like this, there's another property that many systems need: a good sense of balance. When we perturb them or give them a little "kick," they should bounce back rather than fall apart. And in some ways, that's a concept that a famous daredevil named Philippe Petit is quite familiar with.

On the morning of August 7, 1974, Petit was 24 years old, and he was standing on a thin metal wire more than 1,300 feet above New York City, walking a tightrope between the Twin Towers of the World Trade Center. It's hard to say exactly what was going through his head at that moment, but he probably wasn't thinking about biology. Nevertheless, the system in his body that regulates his blood sugar levels has some similarities to the balancing act that allowed him to stay up there in the first place. Both of these processes had to be resilient to disruption in order for Petit to survive.

Tightrope walkers use all sorts of physics tricks to make their jobs easier; for example, they tend to carry a long, flexible pole, which lowers their center of mass and increases the amount of force that would be required to tip them over. But at the heart of it, walking on a tightrope is all about balance: in order to stay on the rope, the walker must stay centered on it. If she finds herself leaning to the right, she has to apply a correction to move back to the center of the rope. If she is buffeted by the wind and pushed to the left, she had better compensate by leaning right, or she'll have about 10 to 15 seconds of free fall to think about her mistake. Balance is all about correcting small perturbations that knock the system off center.

While Petit was wowing the country with his crazy stunt, his body was working hard to make sure that his blood sugar was at a level that would sustain his physical activity. For most healthy people, this level is something around 100 mg/dL of the sugar glucose. Get too far away from this happy center for too long, and many parts of the body, including the brain, stop working. That means unconsciousness, seizures, and even death in extreme cases. Luckily, Petit's blood sugar did not pose a problem.

Regulating the body's blood glucose levels is a difficult job because many events can change those sugar levels. For example, eating a big meal will cause a spike in blood sugar; on the other hand, intense exercise might pull sugar out of the blood to power the muscles. Absent any compensatory or regulatory system, these perturbations would cause blood sugar levels to fluctuate wildly. But because the body does an excellent job of controlling blood sugar levels, people don't drop dead after skipping a meal.

The system that regulates blood sugar basically has two different arms: one that reacts when blood sugar gets too high, and one that leaps into action if the sugar levels get too low. After an influx of sugar into the bloodstream, special "beta cells" in the pancreas release insulin into the blood, which causes cells throughout the body to take up sugar from the blood. Muscle cells, for example, take up the glucose to use as fuel, and the liver stores away excess glucose for later use. These actions lower blood glucose levels back to within the normal range, and this process stops when blood sugar levels are low enough to reduce the amount of insulin the pancreas is releasing. Similarly, when blood sugar drops too low, other cells in the pancreas release glucagon, a hormone that tells the liver to give back the glucose that it stored previously when blood sugar levels were too high. These two opposite mechanisms are the balancing forces that act together to push blood sugar levels back into the normal range whenever they deviate.

In type 1 diabetics, this balancing system is broken because the body's immune system erroneously attacks the cells in the pancreas that release insulin—the beta cells. Without sufficient functional beta cells, the body is unable to decrease blood sugar levels adequately when they get too high. This causes blood glucose levels to fluctuate wildly and often to reach dangerous levels.

Absent this regulation, patients must try to mimic the way beta cells release insulin. Unfortunately, the most common method of insulin delivery for type 1 diabetics is pretty unwieldy. Usually, patients must measure their own blood sugar levels every few hours, and this check often requires a prick of the finger and a portable measuring device.[2] Patients must then self-administer insulin either by injecting themselves with a needle or by manually instructing a pump that is implanted in the skin to deliver insulin.

Even for patients who manage to follow this onerous management procedure, blood glucose levels can still swing outside of the normal range. Patients can't be constantly checking their glucose levels, but that's what normal beta cells are able to do. Even worse, patients must estimate the amount of insulin they need based on the number of calories they expect to get from a meal, which can be difficult to predict accurately.

Recent advances have allowed scientists to build artificial replacements for beta cells that can help keep blood sugar in balance.[3] One example comes from the lab group of Frank Doyle, the dean of the Harvard John A. Paulson School of Engineering and Applied Sciences. The scientists on his team developed a wearable machine that continuously monitors a patient's blood sugar levels and delivers insulin as appropriate.[4] The hope is that pumps like these will allow patients to better control their blood sugar while simultaneously freeing them from the burden of trying to properly dose themselves with insulin. Indeed, one such pump received FDA approval in September 2016.

While scientists can build systems that maintain the blood sugar balance, it is difficult to get those machines to perform as well as the body's own system. The first reason this is difficult is common to both artificial pumps and to the body's normal insulin delivery system: insulin acts relatively slowly—the insulin the pump releases now won't be felt for several minutes. That delay makes it difficult to keep the system stable—it's like trying to guide a car through a high-speed obstacle course by passing notes to the driver from the backseat.

In other ways, a human-designed insulin pump actually faces even more challenges than the beta cells do in a healthy person. While the body can ramp up insulin release in anticipation of a big meal—for example, the parasympathetic nervous system can trigger insulin release in response to the taste of food[5]—a pump that's just watching the patient's blood glucose level can't respond until it sees the sugar actually hit the bloodstream.[6] Even if the pump could predict glucose spikes before they happen, dosing insulin preemptively could be risky: if the pump overestimates the need for insulin, blood glucose levels will crash, which is a much more immediately dangerous situation than having too much glucose.

Still, the pump does a good job of keeping blood glucose within a safe range; it performs significantly better than patients are able to do manually. The pump is personalized for better performance,[7] and it will be especially valuable in people—such as small children—who have trouble controlling their glucose levels with current treatments.

There's another possible solution to glucose management in type 1 diabetics: replacing the insulin-releasing beta cells that the immune system destroyed. Indeed, after many years of research, scientists are now able to make beta cells in the lab. They can take cells isolated from a patient's body, then add many different rounds of chemical signals to prod the cells to change into beta cells.[8] After weeks of treatment, those cells are able to produce insulin. If scientists could find a way to insert these beta cells into the body of a patient with type 1 diabetes and to protect them from the immune system attacks that killed the body's original beta cells, we may eventually be able to restore a patient's blood sugar monitoring system using their own cells.

While human-made insulin pumps have to be very careful about predicting spikes of glucose for fear they'll get it wrong, it is natural for some biological systems to predict the future. For example, the body's circadian rhythm predicts the sunrise and sunset. Ivan Pavlov famously trained dogs to predict that food is coming when a bell rings. Even flinching at a loud noise could be thought of as a prediction of some kind—an evolved response that predicts it's time to duck.

Bacteria, too, try to predict what challenges they'll face next—because they need every advantage they can get. Life is hard for bacteria. They're constantly being eaten by paramecia or worms, or killed off by other microorganisms that don't appreciate the competition for food. Even worse, their environment is constantly trying to kill them. Environmental changes that humans don't even think about require a strong and active response from a bacterium.

As a tiny, single-celled creature, one of a bacterium's primary concerns is keeping its insides inside and making sure that whatever's outside stays there. If a bacterium wanders into a salty environment, for example, it will lose water until it shrivels and dies—unless it actively fights this process. Temperature is also an issue; if a bacterium stumbles into a warm region, all of its proteins—the basic building blocks of the cell—could start to unravel. The life of a bacterium is all about an endless cycle of threat, response, threat, and response in an effort to keep from dying.

But what if the bacterium could predict the future? What if it learned to anticipate the changes in its environment like meteorologists can predict the weather—by noticing patterns in the environment and guessing what comes next? It would finally be able to get out ahead of this threat-response cycle and prepare for the next disaster: it could respond to the next threat before it actually happened. That could help it survive challenges that kill its peers, allowing it to pass along this predictive ability to its offspring.

Indeed, some bacteria seem to have evolved a kind of "memory" of past conditions that allows them to "anticipate" future environmental changes. This memory isn't anything like the way a human remembers something; it's something much more mechanical. Rather, in this context "memory" simply means that the bacterium's behavior at the present moment depends on what has happened to it in the past. This allows it to prepare for upcoming events, but not in a conscious way.

The bacterium *E. coli* faces a predictable environment as it passes through a mammal's intestinal tract. As discussed previously, organisms use various kinds of sugar for energy, but each of those sugars requires different enzymes to help break them down and turn them into usable energy. Animals that eat both milk and starches tend to have the upper part of their digestive tract rich with lactose and the lower part rich with maltose, and bacteria can use this information to their advantage. The bacteria that are traveling through the animal's gut want to eat this food too, so they need to produce enzymes that will help them break down the right sugars at the right time. When *E. coli* encounters lactose, it turns on genes that help it digest lactose, but it also turns on genes that help it digest maltose;[9] this advance preparation presumably helps it to be ready for the change in environment from lactose to maltose that it's likely to see next—and that advance preparation can help it get the most out of the maltose it's about to find.

If the hypothesis is true that the bacteria evolved this behavior to anticipate the appearance of maltose after lactose, we should expect this to be a specific, directional response. Indeed, that is what scientists see: when researchers gave these bacteria lactose, they turned on maltose genes, presumably in anticipation of it coming up next. But it didn't work in the opposite direction: maltose doesn't turn on lactose genes. Since the bacteria run into lactose before maltose, it would not make any sense to prepare for a sugar that they've already passed. And when the researchers repeatedly gave the bacteria lactose without following up with maltose, the bacteria evolved to separate the two responses: the population "learned" through evolution that lactose no longer predicted the appearance of maltose.

One of the most impressive things about biological systems is that they function so well despite how messy they are on a molecular level. The most basic processes that a cell needs to perform—such as making proteins—are subject to considerable fluctuation. If a cell encounters a situation that calls for making ten proteins of a certain type, it's quite likely that it will accidentally make eleven or eight or twenty-five proteins instead. It's like shaking a table covered with pennies and trying to get exactly ten of them to fall off the edge: different numbers of pennies will fall off by random chance, even if the table is shaken in pretty much the same way every time.[10]

The problem is not limited to how accurately the cell can produce a certain quantity of protein within itself; there's a lot of noise in its environment as well. Cells constantly have to react to their environment, and they can do that only through the signals they see—concentrations of small signaling molecules, physical tension on the cell's "skeleton," or any other measurable feature of the environment. If the input to a system that makes a decision about what the cell is supposed to do—grow, move, or produce an enzyme—gets a different input, it might make a totally different decision.

There are some fundamental physical limits on how precise a cell's systems can be. Imagine a gene that controls its own activity through some indirect feedback—like the negative feedback loop of a tryptophan-producing gene that shuts itself off after it has made enough tryptophan. A cell makes copies of this gene as RNA, then that RNA makes proteins. Those proteins then loop back around to eventually regulate the gene's activity through an arbitrarily complicated mechanism. Since that part of the loop could be literally anything, it is drawn as a question mark in a box:

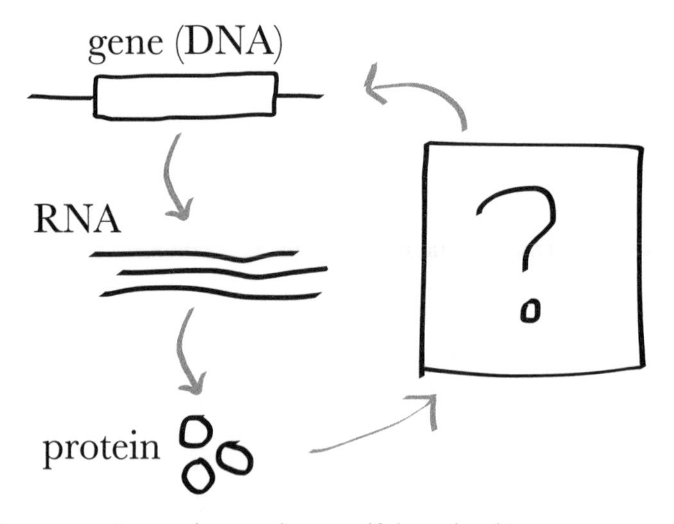

Figure 24: A gene that regulates itself through arbitrary means.

For any system like this, scientists from the University of Cambridge and Harvard University have proved a mathematical limit on how precise the feedback can be—no matter what happens in the box with the question mark![11] The short version of their result is that noise is very difficult to reduce: every time the cell wants to double the precision of this feedback loop, it must spend at least 16 times as much energy as before by making protein 16 times faster.[12] To improve the precision ten-fold, it would have to spend at least 10,000 times as much energy. In other words, attaining high precision in a given process is incredibly expensive for a cell, if not impossible. And while there aren't too many mathematically provable laws in biology, this is one of them. The cell can put anything it wants in the question mark, but it can never do better than this limit.

Despite the inherent noisiness of biological systems, these random fluctuations can't be allowed to derail the system's function, or the organism could die. Somehow, most systems' behavior is robust to fluctuations in their constituent parts.

This precarious balance of noise and information is especially apparent during development. This process will be discussed in more detail in Chapter 7, but for now it's sufficient to say that each of the cells in a developing organism has to decide whether to become a skin cell, or a muscle cell, or a neuron. Individuals of the same species develop into adults that have basically the same body plan—that is, development is pretty reliable—but each individual cell can't see the whole picture. It can only look at the signals that it's getting from its environment and make a decision about what kind of cell to be. The cells in the same positions somehow make the same decisions nearly every time, but they can actually see considerable variation and noise in the signals they receive.

Broadly speaking, in order to make a more accurate decision, the organism needs to expend more energy by making a stronger signal or to have its cells spend a lot more time observing that signal to average out the noise. In the penny scenario, one could make the signal clearer by shaking the table harder and knocking off more pennies: the difference between 1,000 and 1,001 pennies is less important than the difference between 10 and 11 pennies. Or, one could repeat this process many times; sometimes more than 10 pennies will fall, sometimes fewer, but the average will be about right, given enough trials.

But spending lots of either time or energy can be unappealing for a biological system because both are often at a premium. Since every bit of energy the cell uses is a bit of energy it cannot put into finding food and reproducing, it is probably to the organism's advantage to be as efficient with its operations as possible. And if an organism sits around for five hours before it makes every decision, it might not survive to reproduce at all. All life is a delicate balance of energy, time, and information, and evolution has provided some pretty clever solutions to this tradeoff.

Consider a single cell in the middle of a developing fruit fly. In order to make its decision about what kind of cell to be, it can only make use of the limited information available to it. It can measure how much of certain chemical signals are in its immediate surroundings, but it can't see the big picture. If it sees 62 molecules of a "head" signaling protein, should it become a "head" cell or an "abdomen" cell? Is it possible that it just got an abnormally low count of the "head" signaling protein by chance? Since every cell in the embryo is going through this, it's easy to imagine that it might be difficult for the embryo to get development right essentially every time—and the consequences for being wrong could be the death of the whole fly.

Scientists have made calculations of how much information each cell can get from the signaling molecules available to it, and it looks like in some cases the fly may use almost all of the information available to it.[13] If this is true, it means that evolution has managed to balance development right on the narrow line between not having enough information to successfully complete the job and expending too much energy making a signal that is stronger than is necessary. The fly, this would suggest, spends *just enough* energy to make *just strong enough* a signal in order to successfully develop. It seems that though biology is really messy down at the molecular level, there's still enough signal to be pulled out of the noise. And that signal is the information that we and other complex organisms use to go reliably from embryo to adult.

Despite the fact that biology is constantly fighting against noise and randomness, living things occasionally can harness noise to do something useful. Rather than being a nuisance for cells trying to make sense of what's going on in their environments or closely regulating the operation of their internal signaling systems, noise can be good in some circumstances. In fact, some biological systems simply wouldn't work without it.

Noise can be useful for biological systems because it keeps them from being boring—and in the cutthroat world of biology, being boring can sometimes get an entire population killed. For example, in a situation where a microbe doesn't know what's coming next, it might make some sense to diversify: some members of the population will grow quickly, and some will prepare in case disaster strikes. Some cells have evolved to harness noise to produce this kind of diversity in a system's response to a given input.

Imagine that one day a graduate student gets fed up with doing research and decides to get rid of her experiments and move to Tahiti. As she's rushing out the door to catch her flight, she gives her colleague a flask full of bacteria growing in a solution of water and sugar and asks the colleague to get rid of them.

The colleague puts a lot of antibiotics in the flask. "That should do the trick," she thinks. But she soon observes that while most of the bacteria die off, there are a stubborn few—anywhere from 0.01% to 0.0001% of them—that survive. Uh-oh! Those cells are resistant to the antibiotics, right?

Actually, maybe not. When bacteria become resistant to an antibiotic, it means a random mutation changed some part of them so that the antibiotic no longer affects them. For example, penicillin works by inhibiting a protein that helps the bacterium build a protective barrier around itself, so the cells might become resistant to penicillin by producing a different version of this protein that has an altered shape. Penicillin doesn't recognize this altered protein, and the bacteria are unaffected. Alternatively, these cells might develop a modified version of an existing protein that can then be used to break down penicillin. In both of these cases, though, something fundamental has changed about these bacteria: they've acquired a new or altered gene. They have developed some tool that lets them survive the antibiotic.

Once these cells have developed some antibiotic-resistant gene, one would expect them and their offspring to be resistant to future attacks by that same antibiotic since they now have the tools to cope with it. A second dose of the same antibiotic should be ineffective. But when the colleague tries another round of the antibiotic, she once again sees that most of the bacteria die while a few survive. In fact, it seems like these bacteria—the descendants of the survivors—aren't actually behaving any differently from the whole population before the first course of antibiotics.

As it turns out, a few cells survive each round of antibiotics simply because they had been preparing for disaster. In a normal bacterial culture, most of the individual bacteria are going about their usual business of finding food and growing as fast as possible. But every once in a while, a single bacterium will make a choice: instead of growing, it decides to shut down and go into a highly protective state where it happens to be much less susceptible to antibiotics.[14] When the antibiotics arrive, it is these bacteria in the protected state that survive. It's not that they're resistant, it's that they're bracing for impact.

These survivors are known as persister cells, and we can think of this phenomenon as the bacteria hedging their bets at the population level:[15] a small number of them won't grow as fast as they could otherwise, but the population as a whole will be much better equipped to survive a catastrophe that wipes out most of the non-persisters.[16] Indeed, some scientists speculate that persisters might play a role in why certain chronic illnesses are so hard to shake.

Since all of these bacteria are more or less the same, it seems like either all of them should be persisters or none of them should. But in fact, random fluctuations in their cellular components occasionally push an individual bacterium into the persister state. In terms of probability, it's as if each cell randomly flips a coin and becomes a persister cell only if it gets heads nine times in a row. That low probability is just enough to allow a few cells to survive, which can then repopulate after the threat has passed.

Many systems are made up of specialized parts that each has its own particular job, but there are other systems that are made of many essentially identical components. One classic example of such a system is a traffic jam. Sometimes, traffic jams result from an obvious instigating event, such as a collision or road construction. But other times, the congestion seems to simply disappear when one gets to the "front" of the traffic jam, and it's not obvious why there was a traffic jam in the first place. Our intuition is that some idiot at the front of the traffic jam was driving poorly, but it could be that no single person is at fault—everyone is.

Indeed, in computer simulations, realistic-looking traffic patterns can emerge from the collective actions of many virtual drivers. In these models,[17] the drivers follow some very simple rules, for example: (1) don't run into other cars and (2) if you're going less than the speed limit and there's space in front of you, speed up. Those rules sound reasonable, but if everyone behaves this way, traffic jams emerge when there are too many cars on the road. A small variation in the speed of one section of cars or a few people braking suddenly can cause waves of traffic to bunch up behind them.[18]

This kind of emergent group behavior also appears in many biological contexts: fish swim in a coherent school, birds flock away from a predator, and ants forage for food. And just as no single car causes a traffic jam, no single bird or fish is "leading" the group. Instead, each individual animal seems to follow local cues from its neighbors, and the complex behavior emerges from the group. These kinds of phenomena are often called emergent or collective behaviors.

One difficult part of trying to understand emergent behavior is that it is actually relatively easy to build computer models that mimic the behavior of real schools and swarms. For many systems, it's possible to think up multiple models that can explain the behavior we see, and this problem can be particularly acute for collective behavior. Distinguishing among these models requires careful study; it's hard to show that the system is actually following the model's rules rather than some other set of rules that lead to similar overall behaviors.

Despite these challenges, scientists have made progress in understanding some collective behaviors in nature. Consider the search behaviors of ant colonies, for example. When foraging ants successfully return to the nest with food, they seem to stimulate more forager ants to go search for food themselves; similarly, a reduction in the number of foragers returning successfully will tend to cause fewer foragers to head out looking for food.[19] This dynamic adjustment probably helps to ensure that the colonies will intensely harvest food when it is easily available but not expend wasted energy or water searching in vain during leaner times.

The colony's behavior is thought to take into account the costs of foraging. Ants that live in the desert lose significant amounts of water while looking for food, so they must balance the need to find resources with the reality that an unsuccessful search costs precious water.[20] Many of these colonies seem to exhibit conservative behavior as a result: foragers usually only go out in search of food if they smell another forager return successfully.[21] In contrast, ants in other environments, such as the rainforest, may not face the same pressures, and will therefore be more aggressive in their search for resources: foragers in these colonies may head out to look for food even without a positive signal from returning foragers.[22]

Nature's collective systems have some properties in common with the kinds of systems that humans build. The internet—a distributed network of computers—is very good at routing data around damage or network failures, and it uses an algorithm that is somewhat similar to the behavior of the foraging ants in the desert.[23] Professor of biology Deborah Gordon's lab group at Stanford University, which studies these ants, observes a range of foraging strategies in different ant colonies. These different strategies may correspond to different balances between the costs of searching for food and the rewards of finding it—or, analogously, the costs of routing data over the internet. Scientists who study collective behavior hope that the strategies animals have evolved can teach us new methods for solving problems—methods that humans haven't thought of yet.

Other researchers have focused on studying collective behavior by programming simple rules into lots of small, cheap, identical robots.[24] Much like ants or schooling fish, these robots organize themselves as a group without any leader calling the shots. In the future, researchers hope that similar swarms of robots could help search for survivors of natural disasters; go into tiny places larger, smarter robots can't; or mine or forage for resources. Since these robots act as a group, no single robot has to be particularly "smart" on its own. And these kinds of systems could turn out to be the most robust of all: every part would be expendable and replaceable.

PART II
CELLS, ORGANISMS, AND ECOSYSTEMS

CHAPTER FIVE

Beyond Tom Hanks's Nose

Sequencing Technology Is Enabling Scientists to Study All of a Cell's Genes at Once

In April 2013, a 14-year-old boy was treated twice for persistent fever and headaches at a hospital in California.[1] By the time he returned to the hospital in July, he was vomiting, his muscles hurt, and he felt weak. An MRI showed dangerous inflammation in his brain, but no one could figure out the cause.

One likely culprit was an infection of some kind, but if there was indeed an infection, the doctors couldn't find it. None of the tests they ran turned up anything useful, and even taking a small sample of the patient's brain for a biopsy failed to yield any clues. The other possibility was that this was an autoimmune disease, perhaps due to a bone marrow transplant he had received as a child, but there was no way to be sure. Hoping that they were guessing correctly, the doctors put the boy on immune-suppressing steroids. He got worse. The boy developed seizures, which forced his medical team to put him into an induced coma. Doctors had to insert tubes to drain fluid from his head to relieve pressure on the brain.

Finally, the boy's doctors partnered with professor of biochemistry and biophysics Joseph DeRisi and Dr. Charles Chiu, who was at that time a postdoctoral researcher in DeRisi's lab at the University of California, San Francisco. To look for clues, the team examined the DNA in the patient's cerebrospinal fluid (CSF),[2] a clear liquid that circulates around the brain and spinal cord. The vast majority of the DNA they found was human DNA from the patient as expected, but there was also some that shouldn't have been there: DNA from a bacterium called *Leptospira santarosai*. This was the clue the doctors needed; *Leptospira* had infected the boy's brain. The doctors started the patient on a high dose of antibiotics, and he improved quickly. After a short period of rehabilitation, the boy was discharged from the hospital, nearly as healthy as he had been pre-illness.

Even a few years earlier, this happy result would not have been possible. Previous tests on the boy's CSF to identify bacteria had failed, and the positive diagnosis of the boy's *Leptospira* infection relied on being able to identify the DNA found in his CSF and to do it quickly enough to actually impact treatment decisions. To identify the source of the DNA, scientists made use of the fact that DNA is made up of long strings of four fundamental building blocks called bases, commonly represented as A, G, C, and T; determining the string of bases that makes up a DNA molecule is called sequencing the DNA. Since the sequence of bases in DNA varies from species to species, the researchers could identify where the DNA came from based on its sequence. They compared the DNA they found in the patient's CSF with known sequences from millions of possible infectious organisms to look for matches. In this patient's case, scientists had to sequence and analyze billions of bases in order to find the answer—and they did it in about 48 hours.

The ability to sequence that much DNA quickly and cheaply is a relatively new luxury that is the result of enormous improvements in sequencing machines. The progress scientists have made in sequencing is often compared to speed increases in computers, for example. Moore's law famously says that the speed of computers tends to double every year and a half;[3] these enormous performance gains mean that a 2016 iPhone is about 40,000 times more powerful than the computer that landed Apollo 11 on the moon.[4] But sequencing has progressed even more quickly than computers have. In 2003, a large international team of scientists finished a twelve-year, three-billion-dollar project to sequence for the first time the three billion bases of DNA that contain all of a person's genetic information. In 2016, a single scientist could sequence that same amount of DNA in a few days for about a thousand dollars. Sequencing machines the size of a paperweight can now plug into a USB port and show results on a laptop in real time.

Sequencing is a critical tool for biologists in part because it applies universally; every living thing we know of uses DNA and its closely related cousin, RNA.[5] Each of the trillions of cells in the human body, for example, carries a full set of instructions for building all of the parts the cell needs. That set of instructions, encoded in DNA, is called the genome, and those parts are called proteins. Each region of DNA that tells the cell how to make one type of protein is called a gene—a gene is like a DNA blueprint for a protein. To make a protein, the cell first makes a temporary copy of the gene using RNA, then uses that RNA as a template for making the protein. Those proteins then go on to perform pretty much all of the critical functions of the cell: to form the cell's flexible "skeleton," to extract energy from food, to sense conditions in the environment, to talk to other cells, and to perform many other functions.

Sequencing DNA and RNA is important for two big reasons. First, it has given us the full DNA sequence for many organisms, including humans. Since the information that says how to make the cell's proteins is stored in the genome, knowing the DNA sequence tells us the types of parts that a cell has at its disposal. And second, sequencing can tell us which genes are "activated"—making their corresponding proteins—and which are not at any given time. Generally speaking, more temporary RNA copies of a gene means that more of that protein is being produced, so we can take all of those temporary gene copies out of the cell, sequence them to find out how much of each is present, and thereby find out how active each gene is.[6]

Before 2000, the tools researchers had available only allowed them to look at a handful of genes at once. One such tool, called a Northern blot,[7] allowed visualization of a gene's temporary RNA copies as blurry splotches on X-ray film.[8] This procedure allowed researchers to measure the abundance of the RNA by seeing how dark the splotch was, but the method was laborious and often made use of radioactive isotopes to label the RNA of interest—all to determine the activity level of just a few genes.

One of the early pioneers of the field sometimes says that studying one gene is like watching Tom Hanks's nose and expecting to understand *Sleepless in Seattle*.[9] But sequencing now allows scientists to take a snapshot of what all of the genes are doing: we can look at what's happening in the whole cellular system at once.

To understand how knowing the state of all of the genes helps scientists to study cells, imagine a space alien who has just landed on Earth. The alien stumbles across a human village, and she wants to understand how the different villagers work together to keep the town running.

The first thing she realizes is that each person in the village has a specific job. If she kidnaps the carpenter, for example, no furniture gets built, and she can deduce that the carpenter is necessary for building furniture. So one strategy that the alien could take would be to knock out different parts of the system to see what happens. And though this rests on the assumption that there isn't a lot of redundancy—if there were two carpenters in the village, the other might be able to pick up the slack when his counterpart mysteriously disappears—it still provides some pretty valuable information about what functions each part of a system might be involved in. Likewise, if the alien drops her kidnapped carpenter into a village that doesn't have any furniture, it's a pretty big clue if fancy nightstands suddenly start appearing all over the place. In that case, we would say that having the carpenter is sufficient for furniture production to happen. And indeed, this same strategy works just as well for biological systems: a lot of progress in biology has come from making a gene not work properly and observing what happens.

Now imagine the alien has a line of crystal balls that lets her see the status of all of the villagers at once—the village equivalent of sequencing. If the villager is actively doing something, the crystal ball that represents him turns black; if he's inactive, it turns white.

Here's one possible state of the village:

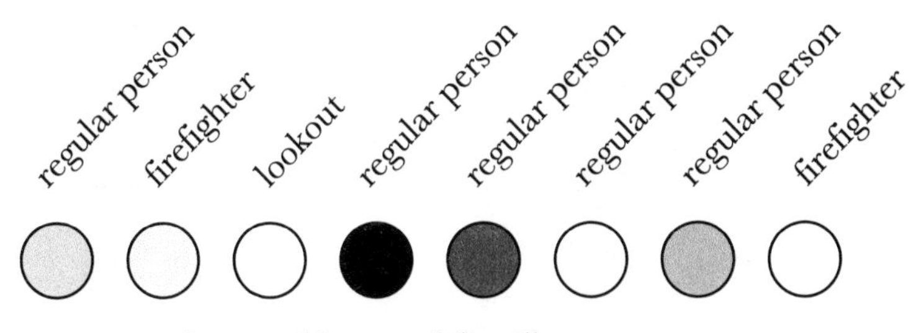

Figure 25: One possible state of the village.

Using this setup, the alien can observe how the community responds to various perturbations to the system. Let's say she wants to study how the village responds to fire, so she sets one of the fields ablaze in the middle of the night. First, the lookout sees the smoke and runs to wake the firefighters, and they immediately get to work putting out the fire. Then the regular townspeople start to notice that something's amiss, and they gradually wake up to see what's happening. Figure 26 shows what the alien would see in her crystal balls over time.

In the first row, only the lookout is active because he's the first to notice the fire. In the next row, the firefighters' columns turn black as the lookout alerts them. And finally, you can see the regular villagers waking up in the later rows. With this grid, the alien can see which parts of the system don't change too much, which change a lot, and which components might make up sub-networks of interest. Depending on her interests, she might then try to figure out exactly how these pieces are connected—how the lookout notices the fire, how he alerts the firefighters, and so on—or to compare the response to fire with the response to flood.

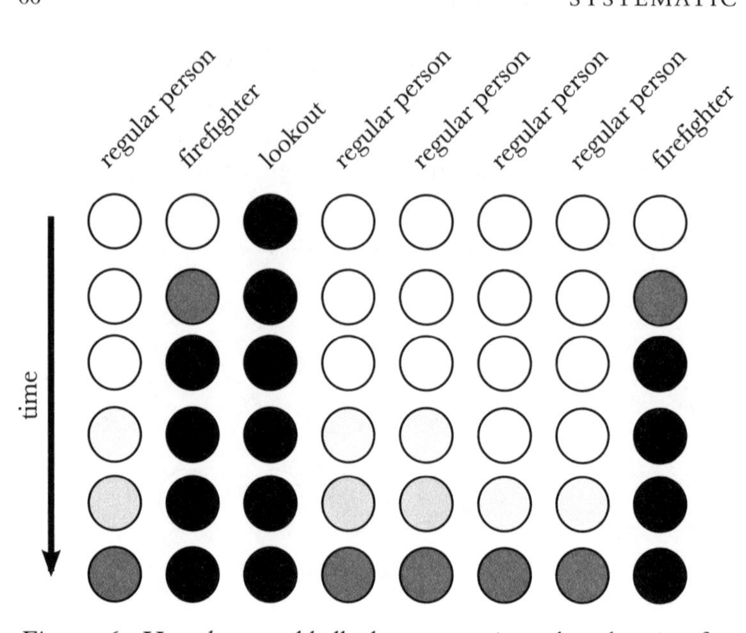

Figure 26: How the crystal balls change over time when there is a fire.

Audrey Gasch, an associate professor of genetics at the University of Wisconsin at Madison, was one of the first generation of scientists to be able to use the biological version of these crystal balls. Today, most experimenters use RNA sequencing when they want to look at all of the genes in a cell, but Gasch actually started out her career using a precursor to RNA sequencing; the function was essentially the same, though. "It allowed us to see, really for the first time, the entire picture," says Gasch. Scientists can study just about any cellular process with sequencing, and Gasch and her lab mates used their newfound abilities to look at how yeast respond to environmental stress[10]—challenges that threaten the yeast cell's very survival.

Several researchers in Gasch's lab at the time studied various types of cellular stress. "One person was really interested in starvation," she recalls; others looked at high temperatures or salty environments that tend to suck the water out of cells. Gasch herself was interested mostly in oxidative stress—when the yeast are exposed to chemicals that can damage their DNA. Though they were all studying different types of stress, a pattern emerged when Gasch and her colleagues compared their results. "We realized that some of the same genes were being triggered by what we thought were totally different conditions," Gasch recalls. It seemed as if the yeast were mounting a common, core response to sub-optimal conditions: there was a set of genes that always seemed to respond in the same way for all of the different types of stress they studied.

And the response was massive. About 350 genes get turned on in response to stress, and almost double that number get turned off. Almost 17% of all of the genes in the yeast cell responded to the stress in some way, according to Gasch. "The breadth of the response and the types of genes that were included was really a surprise."

So what are these genes doing? Gasch's team hypothesized almost immediately that these genes are helping the cell deal with the stress—"but it turns out that that's not true," says Gasch. If the researchers prevented the genes that normally turn on in response to stress from being activated, the cells still did just fine. But these cells did have another problem: while normal cells would be more prepared to handle another dose of stress after the initial exposure, the cells with this stress response blocked couldn't do that. "If we take cells and give them a mild dose of one stress," Gasch notes, "this initial response allows cells to become super-tolerant to a subsequent dose of stress that would otherwise be lethal."

"Cells need to maintain an environment where things work properly," says Gasch, and it seems like a lot of these stress-response genes are involved with keeping everything in order. Some of the genes that are turned on make proteins that help protect cellular structures like DNA, for example, and others are involved in recycling existing cellular parts to help produce more energy that the cell can use to deal with the problem. As for the components that are turned off, many of them are used to make new proteins. One explanation is that the cells might be dialing back on making things that aren't as important for helping to deal with stress; this would hypothetically let them focus their energy on surviving the stress.

Once Gasch's team knew the parts of the cellular system that were involved in the response to stress, they could start trying to figure out how all of those parts fit together. To do so, they used mathematical tools to combine data from many different kinds of experiments to make one best-guess model of the stress response system.[11] This model can then guide future experiments into how this large set of genes helps the cell prepare for future stress. "Without looking at the entire regulatory network in the cell and seeing how that pathway is connected to everything else, we don't understand the system at all," Gasch says.

According to Gasch, understanding a cell's ability to prepare for future stress has many applications, from industrial applications like biofuels to treating human patients undergoing heart surgery or a stroke. "For both strokes and heart surgery, it turns out that much of the cellular damage happens not during the initial event," she says, but because the cells in the brain or the heart go through a period when they're not getting oxygen. When blood flow is restored, "the cells suddenly get a big dose of oxygen, and they have an oxidative stress response," Gasch says. Some experimental medical treatments are now beginning to incorporate these lessons. "Doctors can minimize that damage if, instead of letting the floodgates open and all of the blood flow in a giant dose of oxygen, they give short little bursts of blood flow," she says.

Gasch uses sequencing to study yeast, which are only single cells. On the other end of the scale, multicellular organisms like humans have trillions of cells that all behave rather differently. Cells have "jobs" or specializations that cause them to look and act very differently from one another. A white blood cell and a brain cell have distinct shapes and perform characteristic functions; these properties are caused by different underlying gene activation patterns. For example, a cell that has activated a set of related genes that includes one called *neurogenin 2* is probably a neuron—a brain cell. All of the cells in your body have essentially the same genes, but by activating different genes at different times, cells can produce a huge range of behaviors.

When scientists first started looking at the gene expression (or activation) patterns of different types of cells, they weren't surprised to find that obviously different cell types have different patterns of gene expression. But it was a little more surprising to see that even two cells that look really similar on the surface, such as two breast cancer cells, can have very different patterns of gene activation. First, researchers measured how strongly each of the genes are turned on in many different cell types. Then, using some standard mathematical techniques, they categorized cells based on how similar their gene expression patterns were. That meant cells that both had gene A turned on and gene B turned off tended to be counted as similar, and they were grouped together in this analysis.

When scientists did this for tumor cells from many different types of cancer, each major cancer type, like breast cancer or lung cancer, grouped together. Each different kind of cancer grows in a different tissue, is derived from a different cell type, and needs to do different things in order to escape the natural checks our body has against uncontrolled cell growth—and their patterns of gene activation reflect that. Finer divisions also became obvious. Rather than showing up as one big uniform group, the group of breast cancers actually looked like a few different types of cancer clumped together.[12] These tumors, all historically thought of as just "breast cancer," are actually multiple related diseases. We now know that the same is true for many types of cancer. Cancer cells from two patients—and even two cells within the same patient—can behave differently if they have distinct gene activation patterns.

Doctors now leverage knowledge of this variation among cancers to help guide our treatment of the disease. The subtypes of breast cancer actually do grow at different rates and respond differently to treatment. Looking at which genes a given tumor is using can tell us how aggressive the tumor will be and how we should treat it. For example, one type of breast cancer, called luminal A, is characterized by the activation of a handful of genes with names like *GATA3*, *XBP1*, and *FOXA1*. Luminal A breast cancer tends to have a good prognosis, and we know that it tends to respond well to hormone therapy.[13] Today, doctors routinely measure many different aspects of a cancer to figure out which treatments to use.

Cancer arises as a result of changes—mutations—in a cell's genes that cause it to grow out of control. Scientists can also use DNA sequencing to learn more about the consequences of those mutations. Our knowledge of these subtypes is getting remarkably specific; in acute myeloid leukemia, for example, patients with cancer cells that have a mutation in a gene called *NPM1* but don't have a change in *FLT3* have a better prognosis when they don't have mutations in *IDH1* or *IDH2* than when they do.[14] And that knowledge saves lives.

Every day there are hundreds of thousands of babies born, each of whom developed from a fertilized egg—one single cell.[15] That single cell divided over and over and over, eventually forming all of the hugely varied cells throughout the body, such as brain cells, skin cells, and liver cells. This initial type of cell was special because it was able to turn into any of the various other cell types. Cells that have this ability to give rise to every other different type of specialized cell in the body are known as pluripotent stem cells.[16] Much like we tell children they can grow up to be anything they want, pluripotent stem cells have the potential to take on any specialization. But most of the cells in your adult body aren't like that; a muscle cell is just a muscle cell, and it can't really change into anything else, much like a banker is pretty unlikely to drop everything and become a plumber mid-career. Though they all started from a pluripotent stem cell, all of the cells in your body eventually changed into a specific cell type, and they did so by changing the genes they activated.

Once a cell has specialized, it almost always stays that way. In fact, until relatively recently it seemed that there was no practical way to take a specialized cell such as a skin cell and turn it back into a pluripotent stem cell. But in 2006, Shinya Yamanaka and Kazutoshi Takahashi, researchers at Kyoto University, were able to "reprogram" normal, adult skin cells into pluripotent stem cells. The researchers took skin cells from mice and used viruses to inject extra copies of genes that they thought might tell the cell to become a stem cell. Injecting extra copies of a gene has much the same effect as activating it, so the targeted cells would make a lot of the pro-stem-cell proteins encoded by those genes. Those pro-stem-cell proteins, the researchers hoped, would then coordinate with all of the other parts of the cell to activate the genes that would be appropriate for a stem cell rather than a skin cell.

After painstakingly trying many different candidate genes, Yamanaka and Takahashi eventually came up with a set of just four genes that were sufficient to "reset" regular, adult cells and turn them back into pluripotent stem cells.[17] These genes produce proteins that in turn change the activity of many other genes, thereby remodeling the specialized cell back into a stem cell. Cells that have been reprogrammed in this way are known as induced pluripotent stem cells (iPS cells). Yamanaka won the 2012 Nobel Prize in Physiology or Medicine for these reprogramming efforts, and his work spawned one of today's hottest research topics.

Subsequent work by other researchers has shown that in some cases, we can make cells jump directly to other cell types. For example, scientists at Stanford University showed that they could force skin cells to become neurons by turning on just three genes.[18]

There are also other, more specialized types of stem cell populations that exist naturally in the adult body. These adult stem cells can turn into only a limited set of cell types; for example, blood stem cells, which are mostly located in the bone marrow, can produce white and red blood cells[19] but cannot produce heart cells or lung cells.

Blood stem cells feature in one well-known medical practice: bone marrow transplants. This procedure is used to replace damaged blood stem cells, which can be destroyed as a result of high-dose chemotherapy. In a bone marrow transplant, replacement blood stem cells are delivered directly into the patient's bloodstream via a tube inserted into a large vein; it's similar to getting a blood transfusion. Those blood stem cells then find their way into the bone marrow, where they settle and take over for the patient's own damaged cells. Other treatments that use stem cells, including one that can be used to treat one of the potential complications of bone marrow transplants, are now coming to market or are in clinical trials.

With the advent of iPS cells, scientists hope that we might one day be able to use them to treat patients with their own cells. We can already take a tiny sample of a patient's skin, for example, and turn those skin cells into iPS cells. Scientists can then use those iPS cells to grow insulin-producing beta cells in hopes of curing type 1 diabetes, as discussed in the previous chapter. But the possibilities go far beyond that: iPS cells could hypothetically be used to replace any cell type in the body; that means it might eventually be possible to grow a whole new organ, such as a heart or a kidney, for a patient. Currently, transplant patients have to take medicines to prevent the body from attacking the transplanted organ: since the organ is made from another person's cells, the body recognizes it as foreign. But if the organ were grown from the patient's own cells, the body should tolerate it.

Growing a fully functional heart or other organ is still science fiction, but the technology is moving in that direction. In 2011, scientists created a trachea for an engineering student who lost his windpipe to a golf-ball-size tumor; they used a plastic framework and the student's own cells to make the replacement organ.[20] Scientists have also made progress in growing primitive kidneys,[21] and researchers can create a kidney with rudimentary function if given the scaffold from an existing rat kidney with the cells stripped out.[22]

Perhaps the most futuristic possibility is the idea of 3D printing organs. Imagine if instead of putting down a layer of ink onto paper, a desktop printer instead put down a layer of cells onto a piece of glass. If we then fed the piece of glass through the printer again to add another layer on top, we could start stacking cells on top of one another. And if we did this thousands or millions of times, we might be able to put the right cells in the right arrangement to create an organ. This hypothetical printer could build up cells layer by layer to produce a functional human heart, ready for transplantation.

One method of investigating how all of a cell's genes are connected into a system comes from scientists who work with yeast—the same type of yeast that brews beer. A yeast cell is more or less a sphere about 1/15 of a human hair across. Yet despite their small size, yeast cells are pretty robust—much more so than a car engine, say. Imagine reaching into a running car engine and ripping out a part. Whether you yank out a spark plug, dislodge the timing belt, or rip out the fuel injector, pretty much anything you pick is going to make the engine grind to a halt. But the same isn't true for a yeast cell. Yeast have about 6,000 genes, and only about 20% of those genes are *essential*—the yeast can't live without them. If one were to disrupt any one of the other 80% of the yeast's genes, the yeast could continue living, albeit sometimes at some diminished level of health. This is a great example of the kind of robustness discussed in the last chapter.

Scientists can learn something about how two nonessential genes are connected by looking at what happens when the yeast loses both of those genes. For example, if we were to disrupt two genes that work in a chain, we wouldn't expect much of an additional effect from getting rid of the second gene; losing two genes is no worse than losing one in this situation because the chain is already broken:

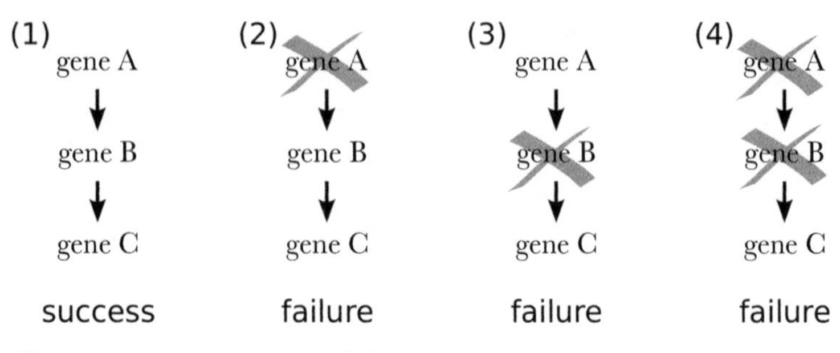

Figure 27: A positive genetic interaction.

Column (1) shows a hypothetical process that happens in a normal yeast cell: the protein product of gene A turns on gene B, which in turn activates gene C. When gene C is turned on, it makes the yeast cell healthier, though its action is not necessary for survival. In columns (2) and (3), getting rid of either gene A or gene B breaks the chain and prevents gene C from being turned on. When genes A and B are both knocked out in column (4), gene C still can't be turned on, but the yeast isn't any worse off than it was without just gene A or just gene B. This is called a positive genetic interaction—when losing two genes is less harmful than expected.

On the other hand, if two genes have redundant functions, one might not see much of an effect by disrupting either one individually, but getting rid of both has a big impact:

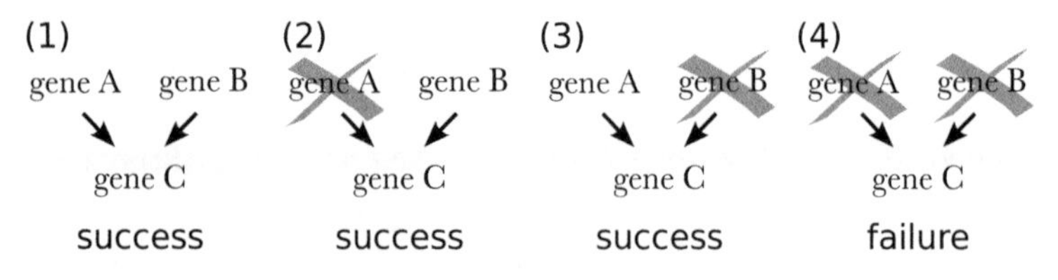

Figure 28: A negative genetic interaction.

Here, either gene A or gene B can activate gene C. Because of this redundancy, losing either gene A in (2) or gene B in (3) doesn't affect the yeast; gene C can still be turned on. But when both gene A and gene B are lost in (4), gene C can't be turned on and the yeast cell suffers. This is a negative genetic interaction: losing both genes together is worse than expected.

Whether it's positive or negative, an interaction between two genes hints that they might be involved in the same processes or do similar things in the cell. That kind of information is useful for building an understanding of how all of the parts of a cell work together, an important goal in systems biology. One step in this direction happened in 2010, when scientists at the University of Toronto and their collaborators published a paper describing the interactions among more than 5 million pairs of yeast genes.[23] This represents a large fraction of all of the possible gene interactions in the yeast, and it provides a starting point for understanding how the cell is put together as a whole.

To visualize these data, the researchers first defined a "profile" for each gene that consists of its interactions with every other gene: perhaps it has a strong positive interaction with genes A and B, a negative interaction with gene C, and no interaction with genes D and E. They then made a visualization using these interaction profiles. In Figure 29, each gene is represented by a circle, and each gene-circle is pulled toward other genes with similar interaction profiles and pushed away from dissimilar genes.[24] At first, it looks like a big, messy jumble.

Even though it looks disordered, there's actually structure in this web of genes. First, notice that there are a few clumps of genes that are all tightly connected together; these show up as dense clumps of dots in the above network. Each of these "modules" seems to be specialized for a particular purpose; most of the genes that perform a similar function tend to show up in the same cluster. For example, in this web, only those genes that are known to be involved in DNA repair[25] are colored in black (see Figure 30).

Most of these DNA repair genes are in one cluster in the lower left; indeed, this cluster mostly consists of genes that are involved in working with DNA. Another cluster has many

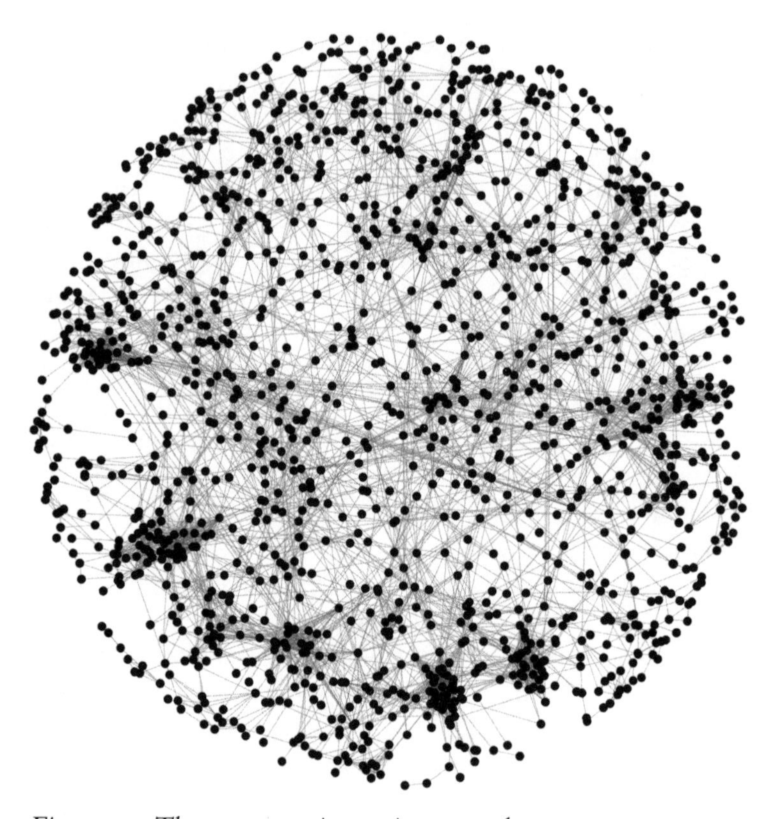

Figure 29: The yeast gene interaction network.

genes that are involved in power generation for the cell, and another has genes that help move proteins around the cell. Scientists can use this web to predict the function of genes that haven't been studied extensively yet; a gene that has a similar interaction pattern to a gene with a known function probably does something similar—guilt by association. And recently, with new technologies for editing DNA, scientists have started doing similar experiments in human cells grown in a dish.[26]

While these types of tools are powerful, looking at all of the genes (or proteins, or any other cellular part) can be overwhelming. If we wanted to use the network of genes to study

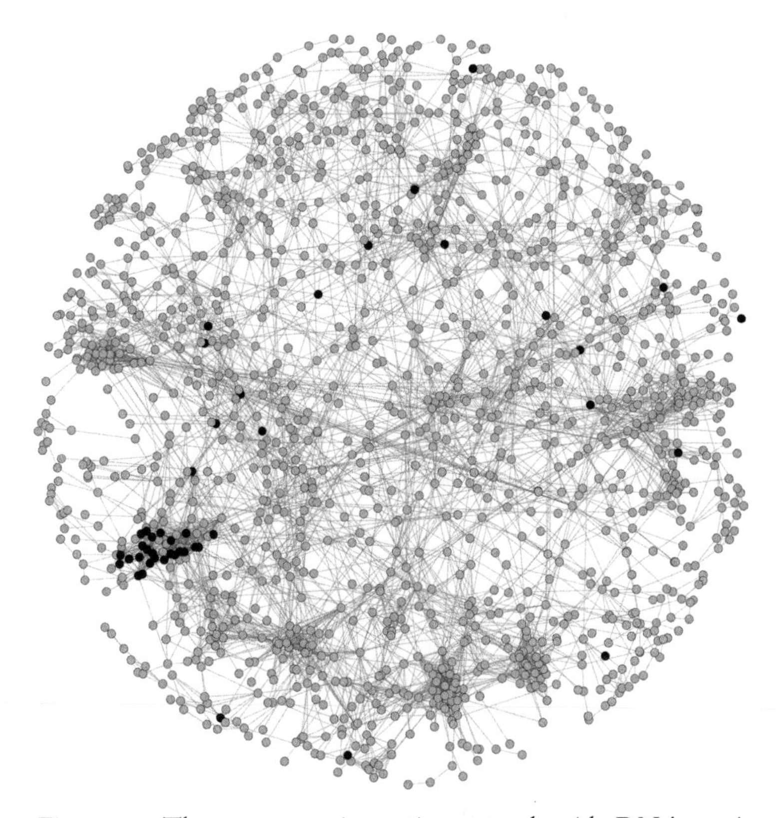

Figure 30: The yeast gene interaction network with DNA repair genes colored in black.

a disease, for example, it's sometimes hard to know where to begin. Some diseases are caused by defects or changes in how certain parts of the network operate, so one would like to be able to figure out which parts of the system to study in detail in order to understand a given disease. In a sense, we would like to build a map of "where a disease is" in the overall system—to figure out which genes are most closely associated with that disease. Since most of the parts of the cellular system are made by genes, all we have to do is figure out which genes can cause the disease when they're disrupted—then we have our clue about which parts of the cellular system to investigate in greater detail.

To build a map of a disease, scientists first needed a way to measure the "distance" between any given gene and the (unknown) genes that contribute to the disease. The answer they came up with was to use landmarks in the genome to help triangulate the disease.

Imagine that a lost hiker wakes up one morning to find himself stranded in the middle of the desert. A helicopter is flying above the desert trying to locate the hiker, but the hiker has only a cell phone to communicate his position.[27] To give the pilot a sense of where he is, the hiker relies on landmarks around him:

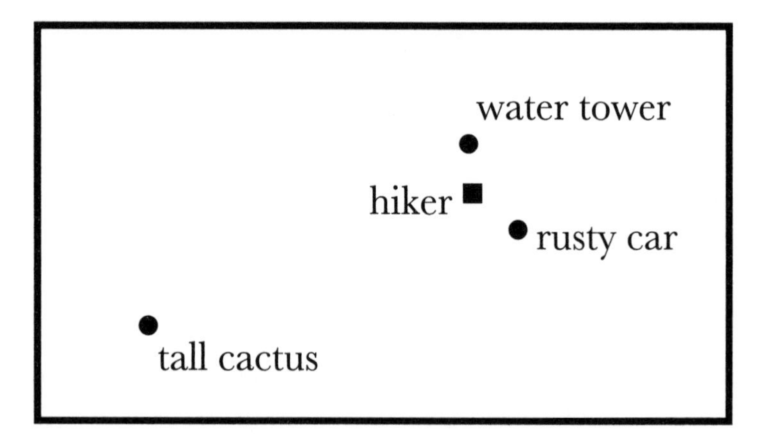

Figure 31: Landmarks around the lost hiker.

The hiker is very close to a run-down water tower and a rusty old car, he tells the pilot, but far away from a tall cactus. The pilot looks around the desert and finds the landmarks the hiker mentioned. The pilot flies to the run-down water tower and the rusty old car, and he finds the hiker waiting, safe and sound. All it took to get the hiker's location was his distance to some known landmarks.

We can use a similar idea to map the locations of diseases in the human genome. Instead of trying to map something in a desert, though, we're trying to map something on a linear strand of DNA. If we were to just take the desert and turn it into a line, it's easy to see how the same strategy of using landmarks would work just fine:

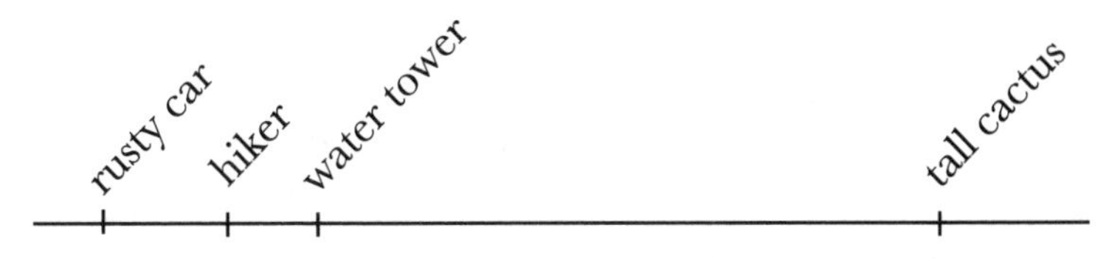

Figure 32: The same landmarks arranged on a line.

But instead of trying to find a hiker, we're looking for the genes that are commonly disrupted in the disease. We don't know which of the many thousand genes those are, though. So imagine that all of the organism's genes and some "landmarks"—common changes to the DNA sequence—are arranged in one long strand of DNA.[28] If we knew which landmarks were "near the disease"—that is, which landmarks were close to the gene that's commonly disrupted in the disease—then we could map the genes that cause the disease onto the strand of DNA just as easily as we found the hiker:

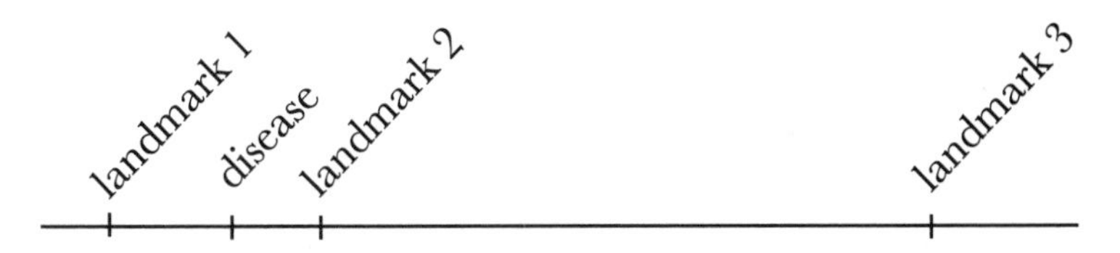

Figure 33: Genetic landmarks arranged on a DNA strand.

The only question then is how to measure the distance between different parts of DNA. Luckily, that distance is hidden in a person's family tree. Whenever humans have children, their DNA is shuffled in a process called recombination. Two genes that are very close together in the genome are unlikely to be separated by this shuffling, but distant landmarks will frequently get split up. So scientists just look for landmarks that seem to be inherited together with the disease under study: those landmarks are likely to be near the gene that's been messed up and is causing the disease.

This strategy can pick out cellular parts that seem to be important for the disease, but it doesn't tell us how those parts behave in the overall cellular system. Still, these types of results have been very valuable: for example, scientists used this technique to find mutations in a gene, breast cancer 1 (*BRCA1*), that can lead to a much higher risk of breast cancer when it gets screwed up. Genetic testing can identify people who carry a risky copy of this gene, and that information can be used to increase vigilance in cancer screening. Combined with a family history of breast and ovarian cancer, this version of *BRCA1* can lead some of those affected—including, famously, the actress Angelina Jolie—to choose a preventive double mastectomy.

A better understanding of disease can also come from using DNA and RNA sequencing. Consider the case of Alexis and Noah Beery, twins who were diagnosed with a rare condition called dopa-responsive dystonia (DRD) at the age of five.[29] For many years, their symptoms were well controlled by the standard treatment for the disorder, supplementation with a molecule called L-dopa; the body can turn L-dopa into the brain signaling chemical dopamine, and it was thought that DRD was caused by a lack of dopamine. But they still had other symptoms, and these got worse in their mid-teens: Noah began to have problems with tremors, and Alexis sometimes had trouble breathing.

Scientists already knew of two genetic mutations that could cause DRD, but neither Noah nor Alexis had either of the known mutations. So the twins' parents and doctors decided to try sequencing the twins' genomes to see if another genetic mutation could explain their symptoms. Scientists from Baylor College of Medicine sequenced the twins and their family to look for clues, and they found genetic mutations that prevented the twins from properly making another brain signaling chemical, serotonin. Supplementation with a chemical that the body can use to make serotonin helped improve both twins' symptoms dramatically.

But both the *BRCA1* and Beery twins cases are limited to a specific change in one gene with fairly straightforward and strong effects. It's easy to predict that a mutation in the gene that makes chemical X could result in the cell not making enough X—and it's easy to guess that adding more X might fix the problem. But we don't yet understand systems of genes well enough to predict the effects of an unstudied mutation in a gene that is connected to many other parts, or to predict the effects of many different mutations scattered around the system. One future direction for systems biology is to understand these network-level effects.

In their efforts to understand the cell, modern researchers have access to more tools and types of global information about the cell than ever before. Besides looking at all of the cell's DNA and RNA, scientists can now, with varying degrees of ease and accuracy, measure vast numbers of proteins, metabolic molecules, and chemical modifications of DNA and proteins. These other types of data can help paint an even more complete picture of what's happening in the cell if used correctly. But more data does not always lead directly to more understanding.

To paraphrase an analogy used by Nathan Kutz, a professor of applied mathematics at the University of Washington,[30] today's scientists—many of whom try to wrangle meaning from large amounts of data—often end up with an answer that resembles the work of the German astronomer Johannes Kepler. In the early seventeenth century, Kepler made a huge discovery using reams of data on planetary motion: the orbits of the planets are elliptical rather than circular. This was a beautiful simplification of earlier models that used awkward patterns of small circles turning on larger circles in order to match the observed motions, and Kepler's model could predict the locations of the planets with exquisite accuracy. Still, Kepler's laws were a description of the motion, not an explanation for it. Even with all of his data and a framework that predicted exactly where the planets would be at any given time, Kepler could never have built a rocket to send astronauts to the moon. For that, he would have needed to understand something about gravity and Newton's laws of motion[31]—the deeper, simpler rules that cause the movement we see in the night sky.

Much like Kepler, scientists today have plenty of data. We can find patterns in the data, and in many cases, we can describe the behaviors we see quite well. The challenge is then to take all of those measurements and use them to infer the underlying systems that govern cellular behavior—to stop describing orbits and start building rockets.

CHAPTER SIX

The Smells of the Father

RNA, DNA Margin Notes, and the Other Missing Parts of the Cellular System

Human DNA is organized into big chunks called chromosomes, and a person's sex is determined by how many of two special types of chromosomes he or she has. If an embryo gets one copy of the chromosome we call X and one copy of the Y chromosome, it will develop as a male, and if it gets two X chromosomes, it will develop as a female. That sounds straightforward, but having an extra copy of a chromosome isn't usually something to be taken lightly.

People who are born with an extra copy of chromosome 21 develop Down syndrome. Fetuses that have an extra copy of most other chromosomes usually don't even survive to birth. But a person with an extra copy of the X chromosome . . . is a woman. Not only is having two copies of the X chromosome not deleterious, it's a normal part of life for half of the population. How can that be?

It turns out that each of a woman's cells has one copy of the X chromosome wrapped up into a little ball of DNA, often called a Barr body. When it's packed up in this way, the cell generally can't make proteins using the balled up chromosome, so females end up with one functional X chromosome per cell, just like males. If it were active, that X chromosome would cause chaos, but by wrapping up the extra X, females are able to inactivate the genes on that chromosome.

Interestingly, the decision of which X chromosome to inactivate seems to be more or less random, so women actually are mosaics of cells: some of their cells express one copy of the X chromosome and the rest express the other. This is how calico cats get their patchy patterning: some of the cat's cells express a gene on one of their X chromosomes that leads to one color, and the other cells express the copy of that gene from the other X chromosome. That's why almost all calico cats are female[1]—male cats have only one X chromosome, so they express it in all of their cells. Even people can be patchy: the mosaic pattern is particularly obvious in women with a mutation in one of the genes on the X chromosome that helps with the development of sweat glands. In that case, the woman will have some patches of skin that sweat and some that don't.

The previous chapter explained that genes contain the information that cells use to make proteins, and proteins do most of what needs doing in the cell; indeed, genes making proteins that in turn modulate the activity of other genes is the core of what makes a cell run. But some of the major components that are responsible for balling up and inactivating the X chromosome are not proteins at all. Instead, the RNA—usually thought of as just a template for proteins—does the work itself. Indeed, one of the major parts responsible for inactivating the X chromosome is a large RNA molecule called X-inactive specific transcript (Xist), which coats the chromosome by spreading out along the DNA.[2] In some types of cells, scientists can actually put the region that makes Xist on an entirely different chromosome and make that one partially inactivate, just like the X chromosome normally would.[3]

More generally, there's plenty going on in cells that isn't just genes making proteins and interacting with other genes.[4] In fact, one of the big surprises that came out of the Human Genome Project was that humans don't have nearly as many genes as scientists had been expecting. In general, the number of genes, and even the total amount of DNA, doesn't seem to have much to do with how complex an organism is. Water fleas have about 31,000 genes. Rice has 51,000 genes. The organism that causes the sexually transmitted infection trichomoniasis has 60,000 genes.[5] And yet humans have only a little more than 20,000 genes—even fewer than a water flea has.

This relative lack of genes was surprising to many scientists when it was discovered. One textbook from 1999 estimated that there would turn out to be 65,000 to 80,000 genes in humans.[6] Since people are so complex, it was hard to imagine us having as few genes as we actually do. On top of this, scientists learned that while DNA is famous for encoding the information that cells need to make proteins, only about 1% of the human genome codes for proteins. The rest of it is used to regulate how those genes behave and to add additional layers of complexity to the cell.

One glimpse of the other parts of the cellular system came when one company tried to make prettier flowers.[7] In the 1980s, Dr. Rich Jorgensen was working at a company called Advanced Genetic Sciences.[8] They were trying to raise more money from investors, so Dr. Jorgensen and his colleagues decided to show off their abilities to genetically engineer plants in a flashy way: they would make a petunia that was *super* purple.

Making a more purple petunia seemed like a no-brainer: they knew which gene was responsible for making the purple pigment, so it seemed like it should just be a matter of adding more copies of that gene into the flower. More copies of the pigment-making gene means more pigment means purpler petunias. Simple.

Except it didn't work out that way at all. When the scientists added more copies of the gene, they didn't get darker-colored flowers, or even flowers that were unchanged. They got white flowers. Adding more copies of the pigment gene seemed to be shutting down pigment production rather than supercharging it.

It turns out that Jorgensen and his colleagues were accidentally activating an ancient viral defense mechanism in the plants. When they added some copies of the gene that makes purple, their actions resembled those of a virus. Viruses work by inserting their genetic material into cells and hijacking them to make more viruses, so Jorgensen's extra copies of the "purple" gene probably looked a bit like a viral attack to the plant. So when the plant found suspicious RNA templates lying around—templates that the plant worried might instruct the cell to make proteins for a virus—it actually used that suspicious RNA to help it find and destroy other templates that look like the suspicious one. When Jorgensen introduced RNA copies that said, "make purple," the cell used those RNA copies to target and destroy all of the "make purple" RNA copies it could find—even the ones that the flower normally makes on its own. With all of the "make purple" RNA copies destroyed, the flower didn't have any pigment and appeared white. This process of turning off genes by introducing foreign RNA is known as RNA interference, or RNAi.

While RNAi might have initially evolved as a defense mechanism against viruses, many organisms actually use this principle to regulate how strongly their own genes are turned on. The cell makes hundreds of short bits of RNA that match up to other genes and target them for destruction: this works in much the same way as the plant cells destroyed the "make purple" RNA copies, but these RNAs are made by the cell itself. These kinds of RNAs provide an additional mechanism that the cell can use to regulate which genes are making proteins and when.

The cell's natural process of gene regulation through RNA-directed silencing can also contribute to various kinds of cancer and other diseases when it goes wrong.[9] An RNA called microRNA-7 appears to help suppress cancer growth and metastasis of certain liver cancers, for example.[10] And the majority of cases of the most common type of adult leukemia have lost, in whole or part, a segment of DNA that allows them to make microRNA-15 or microRNA-16.[11] This inappropriate regulation by RNAs also shows up frequently in immune-related diseases and inflammation.

Building on Jorgensen's happy accident, scientists can also use RNAi as a tool to suppress genes in the laboratory. In the millimeter-long worm *C. elegans*, for example, researchers can easily check the effects of turning off each of the worm's genes; it's as simple as feeding the worms bacteria that carry the right RNA sequence.

Turning on RNAi inside the human body is a bit more difficult—it's hard to get the RNA inside of a person in one piece—but some recent techniques make it more feasible. Since a few human diseases are caused by too much of a single protein, some scientists are trying to use RNAi to treat these types of diseases. For example, a disease called familial amyloidotic polyneuropathy is caused when proteins stick together to form harmful plaques in the nervous system. A therapy that is currently being tested in humans, called patisiran, uses RNA that tells the body to target for destruction the templates that make those proteins, thereby reducing the amount of protein available to form plaques.[12]

And RNAs play many other roles in the cell, too. Besides serving as a template for proteins and helping to degrade other RNAs, they shuttle the building blocks of proteins around the cell. They are a huge part of the cell's protein-making machine, the ribosome. Some can speed up chemical reactions, and some provide a structural backbone for cellular machines. RNAs are so versatile that some scientists even speculate that life may have started out as self-replicating RNA molecules that eventually evolved to include DNA and proteins—and into all life as we know it today.

Even though humans have only 20,000 genes, our cells can actually produce multiple versions of a protein from a single gene. When the cell starts the process of turning on a gene by making an initial RNA copy of it, that copy includes regulatory sequences in addition to the sequences that serve as a template for the protein. Those regulatory regions contain sequences that help decide how the RNA is processed and how quickly that template will be degraded—both of which can affect how and when a protein is made from that RNA.

In order to make a protein, the cell has to process the initial RNA copy of the gene by cutting and pasting the right pieces together to form a final template in a process called splicing. For some genes, there's only one way to put the pieces together; for others, there are dozens of possible final templates. Alternative splicing allows a single gene to make many versions of a protein with many different functions.

One example of alternative splicing takes place in a fruit fly gene called *doublesex*. This gene creates proteins that are partially responsible for determining the fly's sex; one version of the protein turns the fly into a male, and the other turns it into a female. The segments of the RNA that serve as a template for the protein are called exons, and the cell includes or leaves out exons in order to create one of two different possible tails for the protein. If exons 1, 2, 3, and 4 are included, the resulting protein turns the fly into a female, whereas exons 1, 2, 3, 5, and 6 together make the male-promoting protein.[13]

And although it is not common—perhaps because truly severe defects in the splicing machinery are not survivable—problems with RNA splicing play a role in some human diseases.[14] For example, some mutations in the proteins that help carry out RNA splicing can cause retinitis pigmentosa, a disease marked by degeneration of the cells in the retina that detect light.[15]

Alternative splicing can be detected using RNA sequencing, but there are many other important interactions in a cell that aren't necessarily directly reflected in gene expression changes, at least not immediately. Take the modifications that many proteins undergo, for example. Proteins can be chemically modified by attaching small molecules or other small proteins to them. These modifications are endless: proteins can be phosphorylated, acetylated, carbonylated, sumoylated, ubiquitinated, palmitoylated, glycosylated, phosphopantetheinylated, and more. Each of these modifications is used to change the protein's behavior or to signal to other parts of the cell.

Probably the most famous type of modification is phosphorylation, which involves the addition of one phosphorous and three oxygen atoms to a protein. Phosphorylation can affect the protein's shape and thereby change how it interacts with other proteins, and many proteins can be phosphorylated in multiple places. The anti-cancer protein p53, which is activated by DNA damage and other stresses, can be phosphorylated in at least 22 different places[16]—which means the protein could theoretically have over 4 million distinct phosphorylation patterns. And in addition to phosphorylation, p53 is known to undergo at least 8 other types of modification on at least 38 different parts of the protein.[17] All of those modifications help determine which of its many helper proteins p53 attaches to and which actions it takes.

Modifications can also tag proteins for degradation or transport to another part of the cell. Ubiquitin, a small protein, can be attached to other proteins, either singly or in chains of many ubiquitins. A long single chain is usually a signal for that protein to be recycled by the proteasome, a cellular recycling center that breaks the protein back down into its constituent building blocks. Other patterns of ubiquitin modification can cause a protein to interact with other proteins that it doesn't normally attach to. These types of signals are important for recruiting proteins to help repair DNA damage and for responding to some signals from outside the cell.[18] Both phosphorylation and ubiquitin addition are also important for regulating the proteins that govern the cell cycle.[19]

Protein modifications can even be used as sensors of environmental conditions; for example, the protein hypoxia-inducible factor-1α (HIF-1α) seems to use one modification to sense how much oxygen is available to the cell.[20] HIF-1α is modified by a protein called prolyl hydroxylase (PHD) that consumes oxygen in order to attach a pair of atoms—an oxygen and a hydrogen—to HIF-1α. When there's plenty of oxygen around, this modification happens easily, and the modified HIF-1α is tagged with ubiquitin and recycled. But if oxygen is in short supply, the modification of HIF-1α cannot proceed—which means the unmodified HIF-1α is not broken down, and it is free to turn on genes that can help the cell cope with an oxygen deficit.

Finally, cells use networks of proteins that can modify one another in order to regulate several different possible responses to the signals they receive. In yeast, for example, one network of proteins that phosphorylate one another can govern both a yeast's mating behavior and its response to dehydrating environmental conditions.[21] By changing the parts of the proteins that recognize their targets, scientists can even "rewire" these interactions to cause a dehydrating environment to induce mating behavior.[22]

Scientists often think of DNA as a string of As, Gs, Cs, and Ts, so it would be easy to picture DNA as being basically one big alphabet soup. In actuality, the cell's DNA is very well organized. The DNA double helix is spooled around balls, made of proteins called histones, that occur at regular intervals. It's often said that this looks a bit like beads on a string:

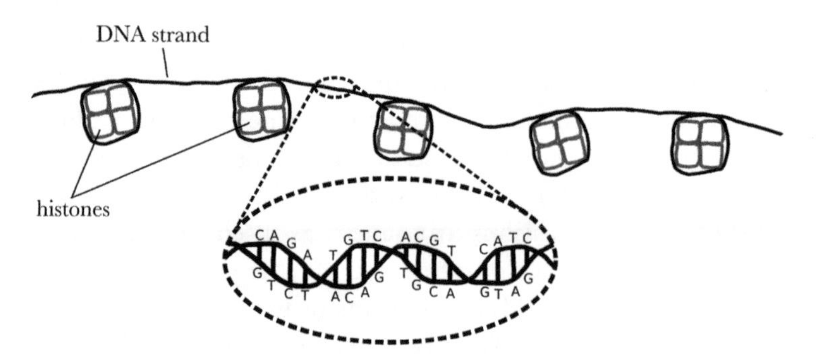

Figure 34: DNA wrapped around histones.

On a larger scale, these beads on a string are then folded up into bigger structures that change depending on what the cell is doing. In some parts of the DNA, these beads and string are wound up and packed tightly together, and in other parts the DNA is loose, allowing free access to the genes encoded therein.

This packing is dynamic, and it is regulated in part by chemical markers on the histones—those little protein balls that the DNA is wrapped around—and on the DNA itself. Those markers change both how the DNA is packaged and how the genes within it are activated. For example, some marks on the histones seem to correspond with nearby genes being active. In contrast, a different kind of mark, called H3K9me3, is associated with tightly packed DNA: since the cell needs to be able to access a gene in order to turn it on, tight winding of a DNA region tends to shut off the genes in that region. The marks on the DNA itself, called methylation, also tend to repress the activity of genes in that region. If the DNA is a recipe for making proteins, changes such as DNA methylation and histone modification are like writing notes in the margins of the cookbook: it doesn't change the printed recipe itself, but it does change how that recipe is used.

Sometimes, these "DNA margin notes" can contribute to certain diseases. Consider a developmental disorder that has its roots in the body's neurons, for example. Human cells, including neurons, have two copies of each gene, one from each parent. For reasons that scientists don't entirely understand, cells throughout the central nervous system preferentially turn off the paternal copy of some genes, in part using DNA margin notes. One such gene is called *UBE3A*, which is involved in adding ubiquitin to other proteins.

Turning off one copy of a gene isn't always a problem, but that means the cell is extra-dependent on the copy that's still active. "It's like a plane with two engines," says Edwin Weeber, director of the Neurobiology of Learning and Memory Laboratory at the University of South Florida: if the plane is flying with two engines and something goes wrong with one of them, the plane can still fly using the other. But with *UBE3A*, the gene that has its paternal copy silenced, "you're already starting off with one engine." That makes everything just a little bit more delicate. "If you lose the mom's gene, that's it—plane's going down."

Unfortunately, in about 1 in 15,000 or 20,000 children, there is a problem with the mom's copy of the *UBE3A* gene, and the child develops a disorder called Angelman syndrome, named after a British doctor, Harry Angelman, who characterized the syndrome in 1965.[23] Weeber lists some of the symptoms: "They never, or very rarely, will say words. Some of them say maybe a handful of words throughout their life." They also often have difficult-to-control seizures. "They walk in a very particular way," Weeber continues, "in that they have a tendency to stay on the tips of their toes when they walk, and they hold their hands out." But despite the host of awful symptoms, there's one bittersweet characteristic of the disability: the children seem to be happy. "Usually when you see an Angelman child, they've got a smile on their face, and they find everything very funny." Even with a severe cognitive disruption, "they are kind of happy children," Weeber says.

But things most people take for granted can be problematic for Angelman syndrome patients. "They almost ubiquitously have a fascination with water—glasses of water, any kind of water at all," and combined with their defects in coordination, this fascination can lead to tragic accidents.[24] The patients' inability to communicate can also be a challenge—"they can't say when they're hurt," he says. "I knew of a child that went into the hospital and they couldn't figure out what was wrong," Weeber recalls. "One bright doctor said, well, have you checked in her mouth? And she had an abscess," he says. "Those are the kind of things that are always a struggle with the kids, just to keep them healthy."

Scientists don't really know why these neurons silence the paternal copy of the *UBE3A* gene, but it appears that the father's copy of *UBE3A* is turned off due to a combination of methylation on the DNA and the suppressing action of a piece of RNA. The paternal copy of that DNA region isn't methylated, which allows the production of a piece of RNA that turns off the father's copy of *UBE3A*. But on the mother's copy, methylation marks on the DNA seem to prevent that piece of RNA from being produced, so the mom's copy of the gene is free to activate. Knowing this, some scientists are investigating whether we might be able to reverse the silencing process for the intact-but-silenced paternal copy of the *UBE3A* gene, thereby hopefully preventing the disorder.

Plants too can fall victim to unfortunate patterns of DNA margin notes. Changes in DNA methylation, for example, appear to be responsible for a deficiency in oil production in the African oil palm *Elaeis guineensis*,[25] the source of most of the world's palm oil. According to the World Wildlife Foundation, palm oil is in "about half of all packaged products sold in the supermarket," including cookies, shampoo, packaged bread, and chocolate;[26] the demand is enormous.

Farmers have discovered that the best oil yields come from hybrid plants that are a cross of two different types of oil palm trees. Among these hybrid plants, though, some plants outperform others, sometimes by as much as 50%.[27] To improve yields even more, scientists have tried to make exact genetic copies—clones—of the best-performing plants; this is functionally equivalent to growing a new houseplant by cutting off a small sprig and planting it in its own pot.

Unfortunately, a large fraction of the cloned plants exhibit an abnormality called mantling that significantly affects oil production.[28] Though these cloned oil palms have the same DNA as the healthy source plant, their fruits are sickly: instead of being round and healthy, mantled oil palm fruit looks a bit like a half-popped kernel of corn. For many years, experimentation reduced the rate of these abnormalities without uncovering the underlying cause of the defects. And despite this improvement, mantling was still a major hassle—especially because it cannot be detected until the plant begins flowing, after growing for several years.

Then, in a 2015 study, an international team of scientists including researchers from the Malaysian Palm Oil Board and Cold Spring Harbor Laboratory found that mantling appears to be caused in part by too little DNA methylation on *Karma*,[29] a type of DNA element known as a retrotransposon. Retrotransposons are selfish DNA elements that can copy themselves into RNA, move to another spot in the genome, and insert themselves back into the DNA sequence in a different place. They're a little like proto-viruses that hitch a ride along with many living things; indeed, almost half of the human genome is derived from transposable elements such as these![30] In oil palms, the *Karma* retrotransposon rides along inside of a gene called *DEF1*, and *Karma* affects how the RNA copy of *DEF1* is spliced. *DEF1* in turn controls the activity of many other genes, and the altered form of *DEF1* caused by a lack of DNA methylation on *Karma* could be the cause of the mantling defect.

While RNA molecules and modifications of the DNA and histones shape the behavior of an individual animal, it has long been thought that these factors are not passed to the next generation—that one's offspring started with a clean slate. That's true in almost all cases, but there are intriguing examples where some nongenetic information can leak through to the next generation in a process called epigenetic inheritance.[31]

One person who is playing a role in uncovering these types of cross-generational effects is Dr. Brian Dias.[32] An assistant professor at Emory University in the Department of Psychiatry, Dias wasn't always the most likely scientist. "While growing up, I was either going to be a Catholic priest, a detective, a radio jockey, or a beach volleyball player," he says. But when he was trying to figure out what to do after college, a conversation with a new professor piqued his interest: "I said, well, let's check this science thing out."

After getting his doctoral degree, Dias took a position as a postdoctoral researcher in the lab of Dr. Kerry Ressler, then a professor at Emory University. There, Dias began working with mice. Scientists commonly teach mice to associate sensory inputs, such as a particular scent, with a negative stimulus, such as a mild electrical shock. They expose the animal to the smell paired with the shock, and after a dozen or so rounds of this smell-shock routine, the mice start to get the idea: this smell means a shock is coming. Previously, Ressler had shown that once the mice are conditioned in this way, they're not only more fearful of the odor—they also grow more of the cells that detect the odor.

Dias wondered: what if those effects could be inherited by the mouse's offspring? He recalls thinking, "This sounds like an outlandish experiment, but maybe we should just try it." He didn't hold high hopes. But Ressler had funding from the Howard Hughes Medical Institute, a generous organization that gives promising scientists money that they can use for potentially crazy projects like this one, so Dias and Ressler gave it a shot.

Dias trained male mice to be sensitive to the smell of a chemical called acetophenone, then mated them with female mice who had never smelled that odor. Then, he looked to see if their kids were also extra sensitive to acetophenone. He remembers the day he got the first results: June 30, 2011. "It was just so striking," he says. The offspring of the mice that had been trained to fear acetophenone had more of the acetophenone-detecting cells—even though they had never encountered that chemical themselves. These offspring are also more sensitive to this odor than other mice are. And this was a specific effect: they were sensitive only to the odor their father had been exposed to. A different odor had no effect.

The increased odor sensitivity of the offspring was weird, but not totally unreasonable. After all, the sperm cells that helped create the offspring were produced from cells in the father that had been around when he was exposed to the odor. Maybe the odor affected the sperm in some unknown way; one wouldn't be surprised to find out that altering an unfertilized egg affected the child that eventually developed from that egg, and the same intuition could hold for the sperm. For this really to be a heritable odor sensitivity, it would have to also be there in the grandchildren of the mouse who was originally exposed to the odor.[33]

So Dias tested the next generation. Neither this group of mice nor their parents had ever been exposed to the smell; yet the mice in the third generation were also more sensitive to the odor. Somehow, these mice had inherited a sensitivity to a smell, not just from their fathers, but from their grandfathers.

But Dias and Ressler weren't done yet. The researchers tried the experiment again, but this time they did something a little different. They again trained male mice to fear acetophenone, but this time they collected the mice's sperm and sent it all the way across campus to a different laboratory. There, someone else fertilized the eggs of a female mouse who had never been exposed to acetophenone. Amazingly, those offspring had more cells that detected acetophenone, even though they had never even been in the same lab as the mouse who was originally exposed to the odor!

The effects reported by Dias and Ressler could operate by several possible mechanisms. Small RNAs that are transferred to the offspring are one possible contender; a self-reinforcing loop where an RNA turns on production of more of itself in each successive generation might do the job, for example. The organism could also sometimes copy histone and DNA modifications from old DNA to new DNA during replication, or it could be something else entirely.[34]

The work being done by Dias and many others on epigenetic inheritance is exciting, but the field is still in its infancy. Many scientists are skeptical of most claims of epigenetic inheritance, especially in mammals. Further research will help clarify the issue.

Epigenetic Inheritance in Humans? Not That We Know Of.

As far as we know, there is no epigenetic inheritance in humans. The possibility is hard to study because people are complicated, and, for ethical reasons, we usually can't do controlled experiments on humans. Another challenge is that people live a long time, so it's hard to watch three generations of people to look for epigenetic inheritance.

A few studies have reported correlations that could hint at such effects in people. For example, some studies have taken advantage of places that keep very good historical records about their citizens to show apparent transgenerational effects. One such study looked at groups born in the late 1800s and early 1900s in a small town in northern Sweden called Överkalix. Scientists from Umeå University in Sweden first tried to estimate variations in food availability based on data about harvests, food prices, and other historical records. They then asked if having more or less food during the critical years of a person's development just before puberty could have effects on the health of their grandchildren. They found that men whose paternal grandfathers had had less food during this part of development died younger than men who had well-fed grandfathers.[35] Later work suggested that deaths from cardiovascular problems and diabetes might also be affected by the nutrition of one's parents and grandparents.[36] And other studies have claimed to find relationships between a person's weight and whether his ancestors smoked as children.[37] These studies have issues that prevent them from being convincing, however. They don't represent the resounding type of evidence scientists would need to declare such an important discovery.

CHAPTER SEVEN

Growing Pains

How Cells and Tissues Coordinate Development, from Egg to Adulthood

Cells were talking to one another long before they were even a part of the same organism. From the time life arose about 3.8 billion years ago until 600 million years ago, there was only primitive, bacteria-like, single-celled life. But then these single cells began to band together and evolve into multicellular organisms—groups of many different cells working together.

There are many reasons why teaming up might be beneficial for cells, but one idea is that they can make better use of communal resources by clumping together. For example, some single-celled organisms break down certain types of food outside the cell, then try to grab the little chunks and bring them inside for further digestion.[1] If a cell is on its own, this isn't very efficient; most of the partially digested chunks float away before the cell can grab them. But if the cells group together, they might be able to collect the food more efficiently. It's like cutting open a bag of feathers in a hurricane. If you're alone, you'll probably be able to grab a few feathers before they blow away. But if you and a few of your friends stand in a circle and open your bags, everyone will be able to grab feathers not only from their own bag, but also some from their friends' bags. Banding together means everyone gets more feathers.

Another big advantage for multicellular organisms is that their cells can have specialized jobs. Instead of a single cell having to take care of all of the functions of life for itself—finding and digesting food, keeping a consistent internal environment, producing offspring—a multicellular organism can divide those responsibilities among many different cells. Skin cells, for example, can focus mostly on protecting the rest of the body from the outside world, and muscle cells can focus on moving. But with this specialization comes a challenge: each cell has to be in the right place and doing the right job. This is the major challenge every multicellular organism faces during its development from a fertilized egg to an adult: to make sure that every cell knows what it should be doing.

Most multicellular organisms, including fruit flies, mice, and humans, start out as a single cell[2]—a fertilized egg. That one cell will eventually divide and grow to produce all of the myriad different types of cells in the adult body. The early embryo is made up of the least specialized cell type, as those cells must eventually produce every other kind of cell. As development progresses and the number of cells in the embryo grows, the cells become increasingly specialized.

The British biologist Conrad Waddington imagined the process of increasing cellular specialization as marbles rolling down a series of branching valleys. At the top, there is only one channel; farther down the hill, the channel splits into many different paths that the marble might take:

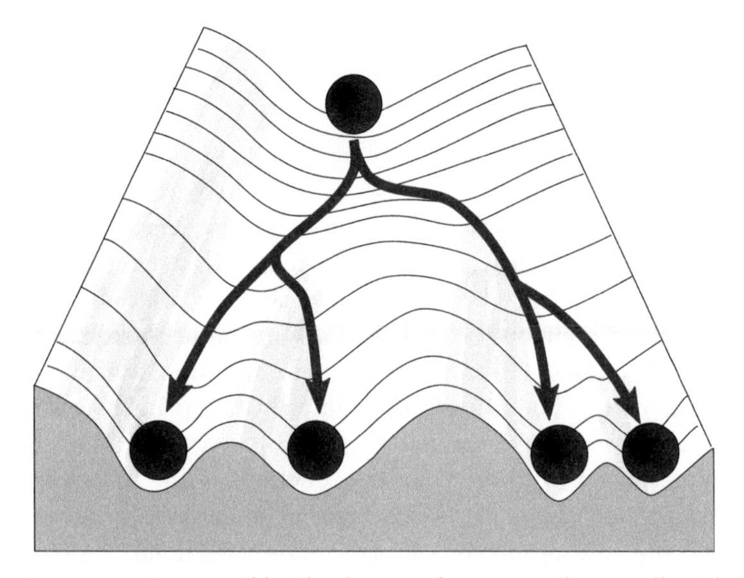

Figure 35: One possible "landscape" that a specializing cell might encounter.[3]

The first branch, for example, might correspond to becoming more like skin and nerve cells (the "outer layer" of the body) or more like muscle and lung cells ("inner" cells).

When the marble comes to a fork in the channel, how does it decide whether to roll left or right—how does it make a decision about which type to become? One common way is using a morphogen—a chemical signal that changes a cell's behavior based on how concentrated that signal is. Much like the smell of freshly baked cookies is strongest right next to the oven, cells that are close to the source of the morphogen will detect a high concentration, but cells that are farther away will see a more diffuse signal. Cells then make choices about how to specialize based on how strong the "cookie smell" is.

This explanation of morphogen behavior is often called the French Flag model, described by Lewis Wolpert in the 1960s. Imagine a blob of cells is trying to create the pattern of the French flag. Cells on the left side need to be turned into blue cells, the cells in the middle should be white, and the cells on the right should be red. To set up this pattern, all we need is a morphogen source on the left side, and we program the cells with the following rules: if there is a high concentration, the cell will turn blue; a medium concentration will cause the cell to turn white; and cells that see a low concentration will turn red:

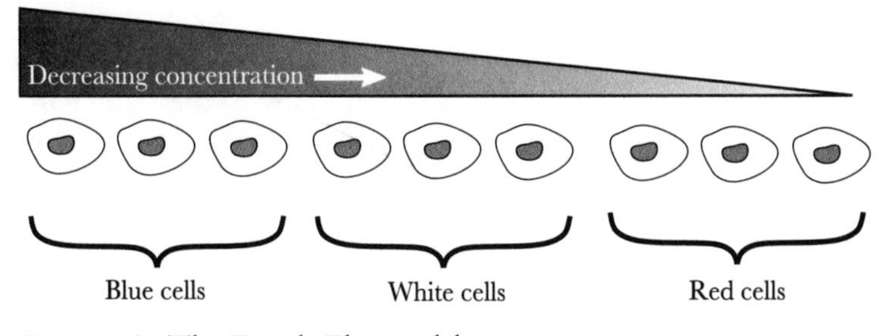

Figure 36: The French Flag model.

And while there is only one decision for a cell to make in the French Flag model—blue vs. white vs. red—an organism can use many different morphogens in combination and in sequence to create very complicated patterns of cell types.

Morphogens literally allow a developing fruit fly to tell its head from its tail, for example. Fruit fly embryos make a morphogen called Bicoid at one end of the embryo, and all of the cells know that Bicoid marks the side that should turn into the head. That "head" signal then diffuses away from the source—which is basically a pile of RNA templates for the Bicoid protein—and forms a gradient in the early embryo. The source is circled on the left side of the diagram below:

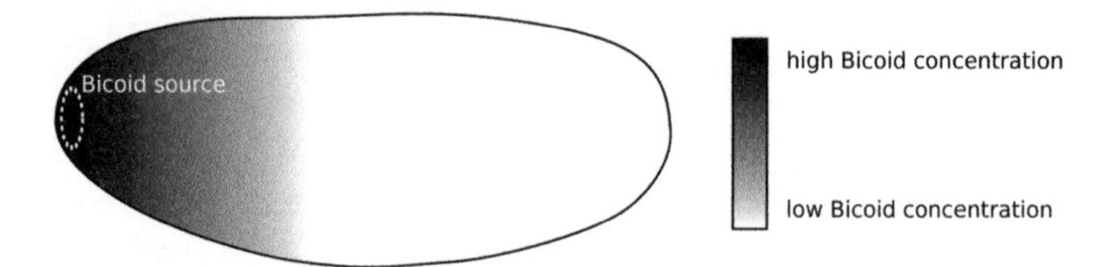

Figure 37: A Bicoid gradient in a fruit fly embryo.

Cells throughout the embryo can then measure how much Bicoid is in their immediate surroundings to know approximately how close to the head they are, and thereby help decide what kind of cell they should become. And that initial location of the Bicoid source in turn came from the mother, who placed it on one end of the egg before the egg was even fertilized.

The process for setting which side of the embryo is the belly and which is the back works on many of the same basic principles: signaling molecules and genes with names like *Toll* and *dorsal* combine to tell each of the cells approximately how far from the back or belly of the embryo they are.

In addition to the head-tail and belly-back axes, there's also a third axis that human and mouse embryos have to worry about: they must tell left from right. Externally, a person's left side looks more or less like a mirror image of her right side, but internally, that's not true. Many of the internal organs are arranged asymmetrically inside of the chest—the heart is on the left side of the body, for example. In mice, this left–right difference seems to be established by the flow of fluid in a cavity in the early embryo. This flow is driven by tiny hair-like structures called cilia that beat in a circular motion to drive the fluid to the left.[4] Other cilia then seem to sense the direction of fluid flow and begin a pattern of signaling that defines which way is left.

Once the embryo knows which way is up, it can start to create ever more complex patterns. In order to do this, the embryo uses patterns of gene activation. Starting with a simple gradient, each round of gene activation makes an increasingly specific pattern and ever-more-specific stripes of gene activation across the embryo. For example, here's an image of a developing fruit fly embryo with a stain that highlights cells that have activated a gene called *runt*:

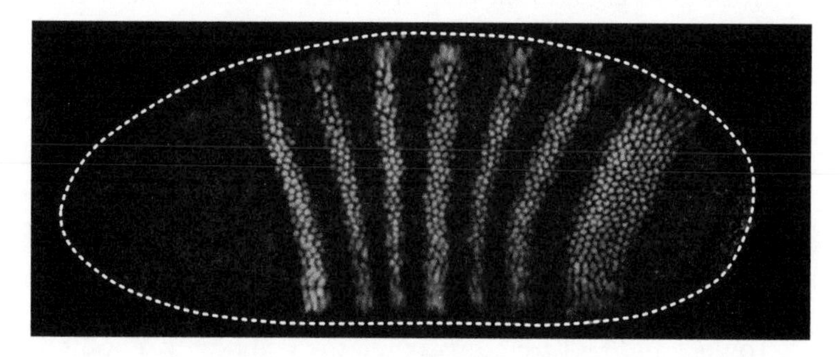

Figure 38: Stripes of cells that have turned on a gene that governs early embryo patterning. The outline of the embryo is shown with a dotted white line.

Each round of stripes sets up the patterns of the ones that follow. The adult fly has many different body segments, and those stripes will eventually help define which cells turn into each segment in the adult fly.

There is also some positional information in a developing embryo that comes from specific cells that help instruct their neighbors what to do. Hilde Mangold and Hans Spemann at the University of Freiburg in Germany uncovered one such source of signaling in the 1920s when they transplanted a clump of cells from one amphibian embryo to another and got an animal with two heads.

The cells of this region, now known as the Spemann-Mangold organizer, release signals that tell the surrounding tissue to become nervous-system-like.[5] When an embryo gets another Spemann-Mangold organizer transplanted into the wrong place, the surrounding tissue doesn't know that something is wrong—it just follows directions. But amazingly, these mixed-up signals still produce recognizable body structures that are simply in the wrong place—which goes to show how robust development really is.

While all of this patterning and signaling is going on, the cells of an embryo are also moving around and sorting themselves into distinct groups. Dr. Philip Townes and Dr. Johannes Holtfreter of the University of Rochester demonstrated one such type of sorting in 1955. The pair of researchers were experimenting with amphibian embryos that had a useful property: when the researchers put the embryos into a high-pH bath, the embryo broke apart into individual cells.[6] Then, by returning the pH to normal, the scientists could reverse the process and make the cells stick to one another again.

Townes and Holtfreter disrupted the embryos at a point in development when the embryos have three major layers of cells: the endoderm, or inner layer; the mesoderm, or middle layer; and the ectoderm, or outer layer. The researchers then tried taking any two of the layers, breaking them into their individual cells, mixing the cells of the two layers together, and allowing them to re-associate.

Initially, the cells just clumped together randomly; the two layers were evenly mixed together. Soon, though, the cells started sorting themselves into the two different layers they had started from. Eventually, the two layers formed once again, and one layer came to surround the other (see Figure 39).

Townes and Holtfreter were particularly struck by the order in which the layers reconstituted themselves. The cells

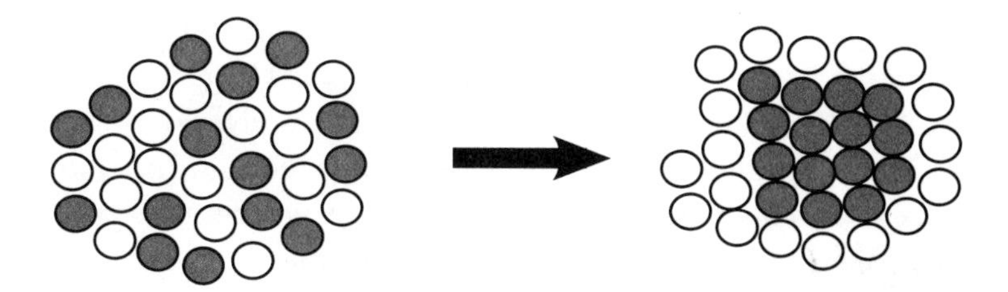

Figure 39: The disrupted cells automatically sort themselves into the correct layers.

that came from the endoderm, the inner layer, always seemed to end up on the inside of the re-formed ball of cells. Cells from the ectoderm, the outer layer, tended to group up on the outside of the ball. And indeed, when the researchers put all three layers together, they reconstituted the layers in the correct order. Somehow, the cells could reconstruct a mass of cells with the same general organization as the original embryo, even after the cells had been split apart completely.

The secret to the cells' sorting ability is that they stick to other cells from the same layer. This stickiness is made possible in large part by the biological equivalent of Velcro, proteins called cadherins (for calcium-dependent adhesion) that sit on the surface of the cell.[7] There are several types of cadherins, and those of a given type tend to stick best to others of the same type. Cells from the same layer tend to have the same type of cadherin, and thus stick well to one another. This stickiness allows the layers to separate out, in much the same way as oil and water don't mix: water molecules stick to other water molecules much more strongly than they do to oil molecules, so the two phases partition. And the ordering of the layers comes from how strongly two cells from the same layer stick to each other. The layer that consists of the cells that stick to one another most strongly ends up on the inside of the ball of cells.

While the cell sorting in reassembled amphibian embryos is passive, other cells undergo epic, active migrations during development. For example, cells from the neural crest, a region of a developing vertebrate embryo, crawl large distances to seed tissues that end up throughout the body. Following chemical signals, they squeeze between other cells and migrate dozens of cell lengths to new homes. In the wake of this great diaspora, neural crest cells go on to become facial cartilage and bones, connective tissue, pigment cells, and some muscles.[8]

Perhaps the most important cellular movement during development is a dramatic event called gastrulation. Before gastrulation, the embryo is more or less a ball of cells, but this cellular migration starts setting up all of the tissue patterns that come later. Indeed, Lewis Wolpert, the inventor of the French Flag model, opined: "It is not birth, marriage, or death, but gastrulation, which is truly the most important time in your life."[9]

The details of the process differ from organism to organism, but the main idea is the same: a portion of the embryo starts to fold inward. That folding forms a pit that is a bit like the hole that would be left if you stuck your finger into a ball of bread dough. In cross section, it looks something like this:

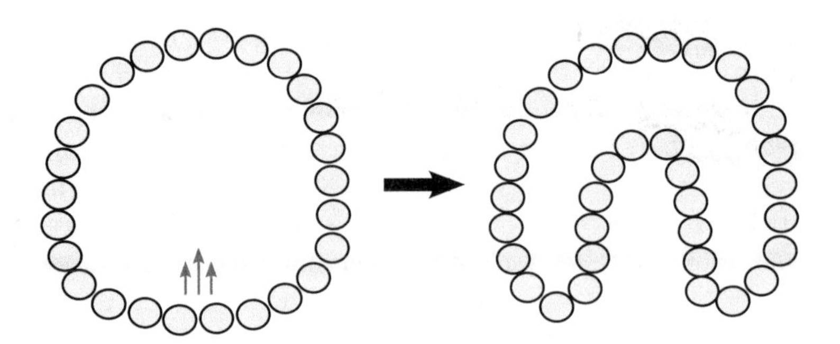

Figure 40: The initial motions of gastrulation.

In humans, that hole eventually develops into the anus, and the top of the fold will eventually fuse with the layer of cells it migrates toward. The hole created by that fusion will become the mouth:

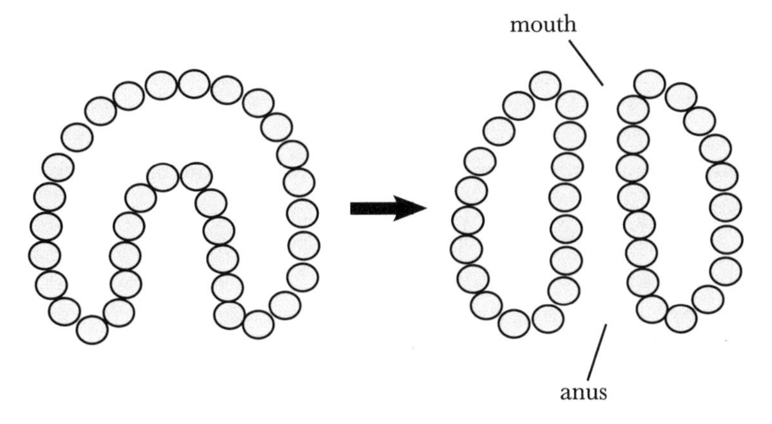

Figure 41: Fusion of the two cell layers creates the mouth.

This major migration coincides with the establishment of the embryo's three main layers—the same layers that Townes and Holtfreter showed could sort themselves. Gastrulation also sets up the "inside" and the "outside" of the body. While a very early embryo is more or less a ball of cells, an adult has the gut, a long tube that goes all the way through the body; it's much like the difference between a basketball and a donut. Having that donut-like hole allows humans to take food in through the mouth and pass the remnants out the other end. Contrast this with simpler organisms, such as jellyfish, which undergo a similar folding process but stop with just a pit. Since it doesn't have a tube that goes all the way through the body, a jellyfish must expel its solid waste back through its mouth. Gastrulation and the formation of the gut tube is what allows humans to avoid that practice.

Once the developing organism starts to take shape, some fine tuning is necessary. While morphogens work very well for making large-scale patterns, such as defining which end of the embryo is the head and which is the tail, making small structures or sharp transitions is harder. If a morphogen comes from a distant source or isn't very abundant, the signals that two adjacent cells receive might not be too different. If one cell has to become a hair follicle and its neighbor a supporting cell based on the morphogen signal alone, it would be hard for the cells to choose their fates reliably; the developing embryo might end up with two hair follicles or two supporting cells.

To help refine the patterns that are established during development, organisms use local, cell-to-cell communication: cells talk directly to their neighbors and come to some mutually exclusive decision, for example. One common way that cells talk to one another locally is using a protein called Notch.[10] Notch sits on the edge of the cell and lets it talk with other cells that are directly touching it. This type of signaling is basically two cells shouting at each other. Imagine both of the cells start out saying "Me me me me me me" at a normal speaking voice to each other. If left unchecked, both cells would get louder and louder until they are shouting, but they'll quiet down a bit if they hear their neighbor yelling. Whether by chance or some other signal, one of those cells will be a little bit louder than the other one to start with, so that will make his partner a bit softer. Now the louder cell keeps getting louder and the softer cell keeps getting softer until the two cells are behaving very differently from each other. They can then use this information—who ends up shouting and who ends up whispering—to make decisions about what kind of cell to be.

Notch can be used to create sharp boundaries, such as the line between a vein (indicated with an arrow) and other cells in a fruit fly's wing (see Figure 42). In this case, Notch and other local interactions between adjacent cells help refine this boundary to create the patterns seen in the picture on page 109.[11]

Notch is also critical in the development of many other complex organisms. In nematodes—little worms—for example, Notch helps make the decision about which cells will become the vulva. In humans, it plays a role in making the liver, the heart, the immune system, the nervous system, and countless other tissues.[12]

Figure 42: A vein on a fruit fly's wing.
(Image modified from Wikimedia Commons[13])

Many of the same signals that help pattern the developing embryo continue to be used in the adult animal. For example, a signal called Wnt—pronounced "wint"—is commonly used throughout development. It helps to define the head-tail axis in many organisms; indeed, Wnt is one of the signals that helps pattern the segments of fruit fly embryos, discussed earlier in this chapter.

Wnt acts in adults in many places, and one of its most famous roles is in regulating the lining of the small intestine. The wall of the small intestine is covered with countless little extruding "fingers," called villi—singular: villus. These villi help the body absorb as many nutrients as possible from the food passing through the intestine by maximizing the surface area of the small intestine. Here's a close-up illustration of one such villus:

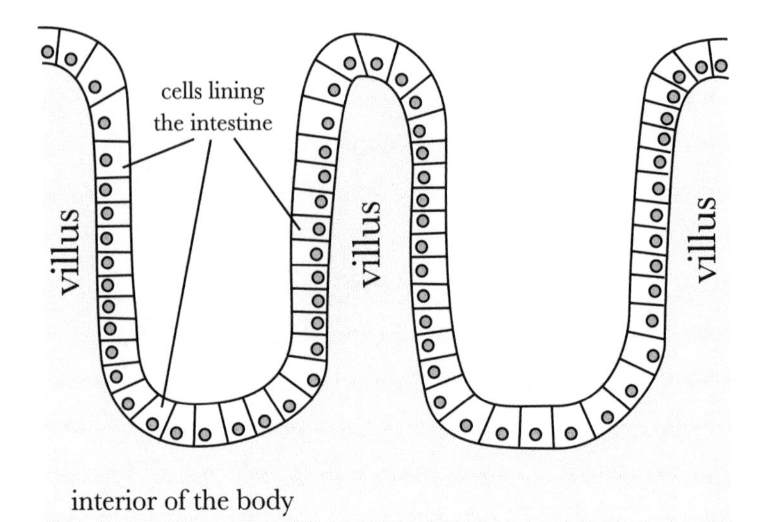

Figure 43: The villi that line the small intestine.

The cells that make up the villi are constantly sloughing off the tip of the villi and being replaced by new cells from the base, a region called the crypt. In the crypt, there are only a few cells that have the ability to divide and make new, replacement cells. These dividing cells need to receive a Wnt signal from their environment in order to retain the ability to divide. Thus, Wnt helps to keep the lining of the intestine in good repair by making sure the replacement cells keep coming.

But even when the signals themselves disappear, the mechanisms at work in development leave lasting marks on the adult animal. For example, vertebrates—animals that have a spinal cord, such as humans—need to form repeating blocks of cells that develop into each of the vertebra. Those vertebra, which you can feel as a series of bumps running down your back, are actually a recording of a clock that ticked in your early proto-spine.

Early in development, the back of a developing human embryo—the part that will eventually become the spine—looks something like this[14]:

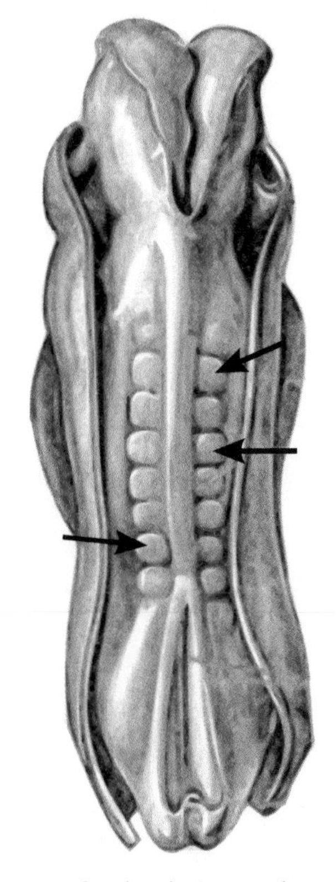

Figure 44: The somites of a developing embryo.

This is a digitized version of a drawing from an old copy of the famous medical textbook *Gray's Anatomy*. Those little balls indicated by the arrows in the center of the diagram are called somites, and those somites eventually turn into the vertebra and ribs. But just before this point in development, the proto-spine is smooth: instead of little balls, there is a smooth tube of cells. The proto-spine has to decide where the boundaries between those repeating somites are going to be.

The prevailing model of this process says that when a human embryo reaches the point in development when it is time to make the spine, each of the cells of the proto-spine has a clock that runs in sync with the rest. When the clock "ticks," a given cell is ready to be a somite—the precursor to a vertebra—and when it "tocks," the given cell is ready to be a boundary between somites. The cells then "freeze" in a wave from the top of the spine to the bottom. Whatever state the cell freezes in—a ready-to-be-a-somite state or a ready-to-be-a-boundary state—that's how it stays. Those repeating somites running down your back then eventually turned into your vertebra—and into a reminder of the developmental processes that shaped you.

CHAPTER EIGHT

No Organism Is an Island

The Interactions Between Individuals and Species that Shape Ecosystems

When the Hawaiian bobtail squid wants to camouflage itself, it relies on some help from another species—specifically, a type of bacteria called *Aliivibrio fischeri*. As the nocturnal squid cruises the waters of the Pacific Ocean, it maintains a colony of *A. fischeri* in a special pouch inside its body. Those bacteria glow with a pale blue light, and the squid uses flaps on its underside to control the amount of light it emits. By matching the light it lets out of the pouch to the brightness of the moonlight shining on its back, the squid avoids casting a shadow that would betray its presence.[1]

This is a productive partnership for both species. The squid produces a nice and cozy environment for the bacterium to grow in, and in return, the bacterium produces light for the squid. This cooperative behavior, called symbiosis, allows both species to thrive.

A. fischeri isn't one to waste energy, though. While the bacterium has the ability to produce light, it will do so only if there are enough other bacteria of the same kind around that can light up at the same time. A single bacterium that starts to glow all on its own produces a weak light and wastes its energy, so this coordination between individual bacteria is just as essential for its success as is its partnership with the squid.

To tell if it is alone or in a group, the bacterium relies on a process called quorum sensing. Each bacterium is constantly giving off a chemical that tells other bacteria of the same type, "Hey, I'm here." This chemical, called an autoinducer, quickly floats away and stays at undetectable levels if the bacterium is alone. But if *A. fischeri* senses a lot of the autoinducer in the environment, that means there are a lot of other bacteria nearby to help with the light show. Confident that its efforts won't be wasted, *A. fischeri* begins to glow.

In order to ensure that all of the individual bacteria start to glow at the same time, the bacteria employ a feedback loop like those discussed in Chapter 2. Each bacterium is constantly producing a low level of the autoinducer, but if it starts to detect the presence of other bacteria, it will supercharge its own autoinducer production. These signals quickly build on one another to cause a rapid rise in the autoinducer concentration; there's no ambiguity about the signal because it goes from low to high quickly.

This communication between individual bacteria is also important for coordinating more sinister behaviors, such as attacking a host. For example, consider *Staphylococcus aureus*, sometimes just called Staph. This bacteria is responsible for tens of thousands of infections in the United States, and when inside a host, it must choose when to attack. From Staph's perspective, the problem is that it is puny compared to the much larger organism that it is infecting. That's why Staph, like many other virulent bacteria, often bides its time until it has enough friends to mount a successful assault on the host.[2]

Staph uses a quorum-sensing mechanism to coordinate its attack and synchronize its virulent behaviors. It produces an autoinducer, a protein that the bacterium releases into the environment to signal its presence. Accumulation of the autoinducer caused by large numbers of the bacteria stimulates the production of toxins designed to hurt the host. Scientists hope that by disrupting this communication—perhaps by blocking the autoinducer signal—we might be able to develop new kinds of treatments for diseases that rely on this communication.[3] To fight against Staph, some scientists have worked toward one day training the immune system to target these bacterium-to-bacterium signals directly[4]—which would perhaps slow Staph's assault or prevent it altogether.

In other bacterial species, quorum sensing is used to coordinate spore production and to govern when they should produce antibiotics to kill off competing species of bacteria. Quorum sensing even helps produce dental plaque; the bacteria in the mouth coordinate to cooperatively manufacture a bacteria-friendly environment, called a biofilm, that coats the surface of the teeth. And while biofilms are pleasant homes for bacteria, they contribute to the formation of cavities and gum disease: removing biofilms is a major reason for brushing teeth.

In more complex organisms, we call the chemicals that one individual gives off to influence the behavior of another individual "pheromones." Pheromones are used for all kinds of communication: butterflies use them to signal a willingness to mate; aphids use them to sound the alarm when attacked,[5] causing other nearby aphids to run away; and dogs and cats use pheromones in their urine to mark their territory. For mice, urine-carried pheromones might play an additional role: these pheromones might be used to identify relatives and avoid inbreeding.[6]

The ant foraging behavior discussed in Chapter 4 is also governed by pheromones in many species. If the ants find a good stash, returning ants use pheromones to lay down chemical tracks that their outgoing colony-mates can follow to the food find.

Indeed, humans can manipulate ants' pheromone tracks for our own purposes. For example, the Nobel Prize–winning physicist Richard Feynman told a story in his autobiography—in a chapter about his natural curiosity—about keeping ants away from his food in his dorm when he was a graduate student at Princeton University.[7] Instead of killing the ants, Feynman says he experimented with them. The solution he settled on involved picking them up as they entered his window and redirecting them to a small pile of sugar. When they returned to the nest, these ants left a trail of pheromone leading to the sugar pile rather than to the pantry. People can also hijack pheromones in less benevolent ways: many home gardeners use traps laced with sex pheromones to attract and kill Japanese beetles, a North American garden pest.

Occasionally, signals meant for one animal can accidentally affect other species. For example, humans produce the hormone estrogen naturally. Women also ingest it in the form of the birth control pill. Estrogen then makes its way through urine into our waste water. When this waste water is not specifically treated to remove such chemicals, estrogen is released into the environment and can wreak havoc with fish populations.

In one example, scientists from Canada and the United States Environmental Protection Agency studied a population of fathead minnows living in a lake in northwestern Ontario, Canada. This lake, known as Lake 260, is part of a protected area set aside for scientific experimentation on freshwater environments. During a study that lasted seven years, researchers introduced tiny amounts of estrogen—a few parts per trillion—into Lake 260, bringing it to estrogen levels similar to what is observed in waters near waste treatment plants.

The effects on the fathead minnow population in the experimental lake were dramatic. Male fish that were exposed to this estrogen-laced water became feminized and expressed high levels of a protein called VTG that is normally associated with egg development in female fish. The male fish even developed early-stage eggs in their testes.[8] As a result of these perturbations, the minnow population crashed. Less drastic but still-troubling effects have been observed in human populations; chemicals that disrupt the endocrine system can alter the timing of puberty, among other effects.[9]

In other cases, the interactions between two species are purposeful and evolved. Consider the bacteria in the soil, for example. Even within a single speck of dirt, there are billions of microbes, and many of them release chemicals, food, or antibiotics that affect the growth of their neighbors.

To get a look at the types of interactions at play in natural bacterial communities, scientists isolated 64 strains of bacteria from a few flecks of soil and tested how they affected one another.[10] Each bacterial strain was first grown in a well of a lab dish on top of a fine mesh; the mesh was small enough that the bacteria themselves couldn't pass through it, but small chemicals or other signals that they excrete could. The mesh was then removed, taking the bacteria with it but leaving behind the signals. This first bacterial strain is called the "sender" because it provides the signals that are left in the well. The scientists then put a different bacterial strain, the "receiver," into the well and monitored how it grew. Fast growth indicated that the sender produced factors that favored the receiver, whereas slower growth might indicate that the sender species produced an antibiotic or other malicious signal.

When the researchers looked at how each species interacted with all of the others, they noticed that there were a wide range of interactions. Some senders strongly hurt the growth of all other species, probably through the production of an antibiotic. Other senders hurt some species while promoting the growth of others; these strains seem to have strong preferences about who their roommates are. But overall, the bacteria in this soil appeared to observe a bacterial version of the Code of Hammurabi: an eye for an eye. This bacterial community evolved lots of reciprocity, where two strains tend to either mutually inhibit or mutually promote each other.

Animals living in the same ecosystem can also interact in passive ways, such as by catching diseases from one another. Consider the great gerbil, a rodent that lives in burrows in the deserts of central Asia. Only some of the burrows are occupied at any given time, but they absolutely cover the landscape in some places. For example, here's a satellite image of a desert in Kazakhstan; each of those little light dots is a gerbil burrow about 50 feet or more across:

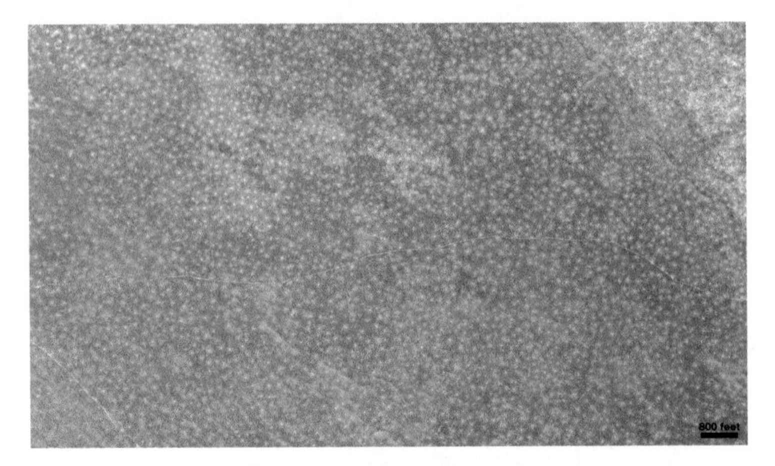

Figure 45: Gerbil burrows in the Kazakhstan desert. (Map data are provided courtesy of Google and Digital Globe.[11])

The great gerbils of Kazakhstan can get plague—the same bacteria that caused the Black Death in the fourteenth century—and waves of the disease occasionally spread through these populations. In 2004, an international team of researchers from Denmark, Belgium, the United Kingdom, Kazakhstan, and Norway noted that plague seemed to spread much more easily through the population of gerbils when the population density was above a certain threshold[12]—when 47% or more of the burrows were occupied. Below this threshold, plague was almost never detected.

Plague is commonly transferred from gerbil to gerbil by fleas. Because of their close living conditions, plague passes easily from one gerbil to another within a burrow, but transmission between burrows is a little more difficult. Thus, whether or not plague spreads throughout the gerbil population depends on how effectively it can hop from burrow to burrow—and that is related to a physics problem: the question of percolation.[13]

Imagine water seeping through a porous rock—a rock that has many tiny holes in it. If there are only a few holes placed randomly throughout the rock, it's unlikely that water will have a clear path from one side of the rock to the other:

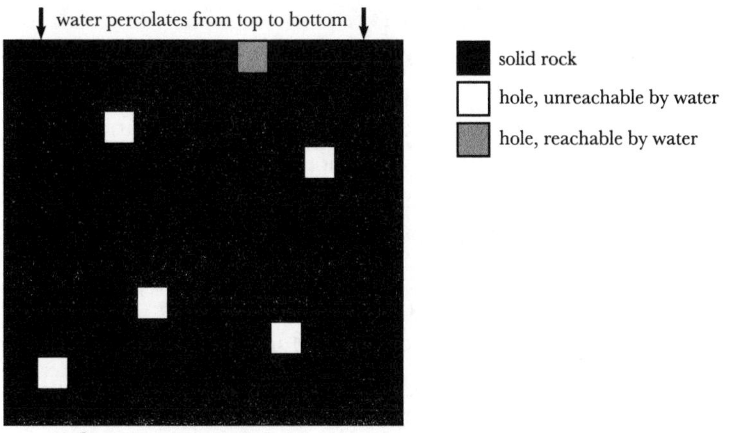

Figure 46: A rock with few holes does not allow water to seep through.

On the other hand, a rock with many holes in it is very likely to let the water pass through because there will almost always be a clear path from one side to the other:

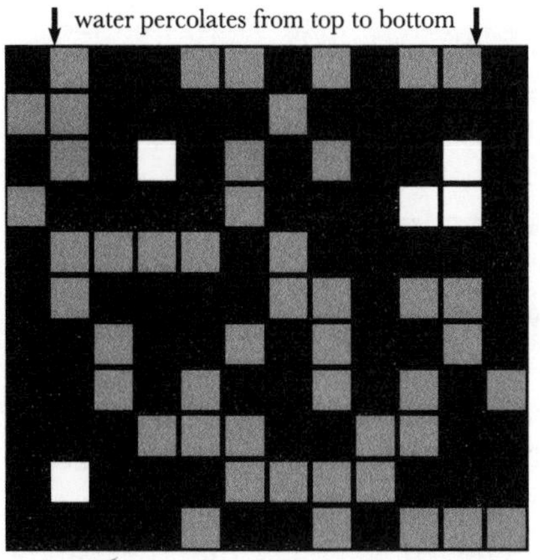

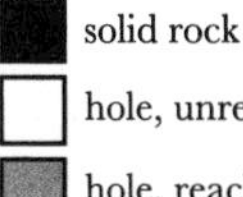

Figure 47: A rock with many holes is likely to allow water to seep through.

Scientists and mathematicians have shown that whether the water can pass through the rock depends on the fraction of the rock that is made up of holes, and there is a sharp transition from "water almost never percolates" to "water almost always percolates" at a threshold fraction of holes. The exact value of the threshold depends on how all of the sites are connected together; for example, in the diagram on the previous page, the threshold depends on whether water can spread on diagonals or just in the four cardinal directions. But regardless of the details, the threshold represents a sharp transition from never percolating to always percolating.

In the case of the plague-ridden gerbils, one can think of an occupied burrow as a hole and an unoccupied burrow as a solid part of the rock. Imagine that plague can be transmitted only between adjacent burrows; in that case, plague is like the water—it can spread through the gerbil population only if there are enough occupied burrows.[14] This type of model could explain why plague spreads widely only when enough burrows are occupied.

In other circumstances, the interloper that spreads rapidly across an ecosystem is not a disease but an invasive species. When a nonnative species invades new territory, it can perturb the native species by competing with them for resources, by acting as a predator, or through other means: the introduction of the nonnative broad-leaved paperbark tree to Florida increased the incidence of fires in the area, for example.[15]

Scientists may have a way of eliminating some invasive species—if society decides it is ethical and wise to use such a tool, that is. The potential strategy rests on the idea of "gene drives"—a sequence of DNA that scientists could design to spread very rapidly throughout a population. Most complex organisms have two copies of each gene, and the parent randomly chooses one of the two copies to pass on to each of his or her offspring. A gene drive makes this process nonrandom; a gene drive makes sure that it is the one chosen to be passed to the offspring. Since all of the animal's offspring get the gene drive, the drive can spread very quickly through a population.[16]

There are many different ways one might try to build a gene drive, but the topic is currently popular because scientists can now do it fairly easily using a new DNA-editing technology called CRISPR.[17] Derived from a primitive bacterial immune system, CRISPR allows scientists to easily program a cell to modify certain sequences of DNA. A CRISPR-based gene drive actually attacks the other copy of the gene and inserts itself in place of that copy.

The appeal of gene drives for dealing with an invasive species is that the gene drive can also carry a "payload"—an arbitrary sequence of DNA. Scientists could someday use a gene drive to remove an invasive plant or animal species by introducing a gene drive into the population that carries a detrimental bit of DNA—for example, a gene that causes all of the offspring to be male, thereby ultimately eliminating the population, or a sequence of DNA that produces proteins that cause infertility. Such a payload would cause an invasive species to die off within a dozen generations.

Gene drives could also be used to solve other problems, such as the burden of malaria infection. Scientists can genetically engineer mosquitoes to be resistant to malaria,[18] so one could put the bit of DNA responsible for that resistance on a gene drive and allow it to spread throughout all of the mosquitoes in Africa. Scientists have already tested the technology to do this in the lab, and it appears to be technically feasible.[19]

But since most gene drive designs would spread freely once released, using one in the wild could have unintended consequences. Because of the potential impacts of the technology, caution is appropriate, and the idea of actually using gene drives in nature is on hold for now. Current laboratory gene drives either are designed not to spread outside of the lab or are otherwise kept safely contained. And the scientists who work on gene drives have wisely begun discussing the ethics of this technology early on in the process—before anyone even thinks about using one.

The most famous mathematical model of interactions between organisms was developed independently by two mathematicians in the 1910s and 1920s. The first, a mathematician named Alfred Lotka, was working on a technique that was a precursor to some of the mathematical models scientists use today.[20] Lotka's technique involved describing a system using a set of differential equations—equations that describe how quickly something is growing or shrinking. A differential equation to describe a population of animals, for example, might say that a population grows proportionally with the size of that population: that is, the number of babies born depends on how many adults are available to have kids. By solving the equations, Lotka hoped to describe the interactions between predators and prey, among other problems in ecology.

To promote his method, Lotka highlighted strange observations about natural systems that he believed his equations could help explain.[21] Lotka was introduced to one such observation, large oscillations in the wild populations of parasites and their hosts, through the work of William Robin Thompson, a researcher in France at the European Parasite Laboratory (later the European Biological Control Laboratory). Thompson used a simple mathematical framework to explain fluctuations in the populations of a parasitic insect, the corn borer, and its host crops. Lotka applied his own equations to this problem and was able to reproduce the oscillatory behavior.

Another strange observation had to do with the then-recently-discovered fact that malaria was carried by mosquitoes. Some scientists advocated for reducing mosquito populations in order to combat malaria, but there was doubt about whether this would work. Most scientists at the time believed that there was not a clear relationship between the number of mosquitoes in an area and the number of malaria cases—which seemed to suggest that reducing mosquito populations wouldn't help the malaria problem. Lotka picked up on earlier work by Sir Ronald Ross, a 1902 Nobel Prize winner, that suggested malaria's spread was not directly proportional to the mosquito population; instead, it spread only when the mosquito population was above a certain threshold. This was suggested as a possible explanation for the apparent lack of a relationship.[22] Lotka confirmed and expanded Ross's work using Lotka's model.

Despite Lotka's efforts to publicize his work, the method failed to attract widespread attention. But around the same time, a well-known Italian mathematician named Vito Volterra was having more success. Volterra became interested in describing the interactions between predator and prey because of work by his future son-in-law, Umberto D'Ancona. D'Ancona was a marine biologist who studied fish populations in the Adriatic Sea, and he noticed that the population of predatory fish had spiked during the years of World War One, when fishing had mostly stopped. Volterra became interested in the dynamics of this fish population and developed a mathematical model to describe it. He ended up with a model very similar to Lotka's, but he had more success at publicizing the results. This model eventually became known as the Lotka-Volterra equations, or simply the predator-prey equations.

A simple version of Lotka and Volterra's model considers two species as the predator and the prey—foxes and rabbits, say. The model says that rabbits make more rabbits, and they are eaten by foxes at a rate proportional to the size of both populations: this roughly corresponds to a fox eating a rabbit when they randomly bump into each other. The fox population grows when they eat plenty of rabbits, but shrinks otherwise.[23]

Given this model, Lotka and Volterra's equations predict that the population of both the rabbits and foxes will oscillate. One possible solution is plotted on the next page. The scales on the X and Y axes are arbitrary—they depend on the exact details of how quickly the rabbits and foxes grow, and on how strong the interactions between them are—but the general dynamics are the same for almost all choices of parameters:

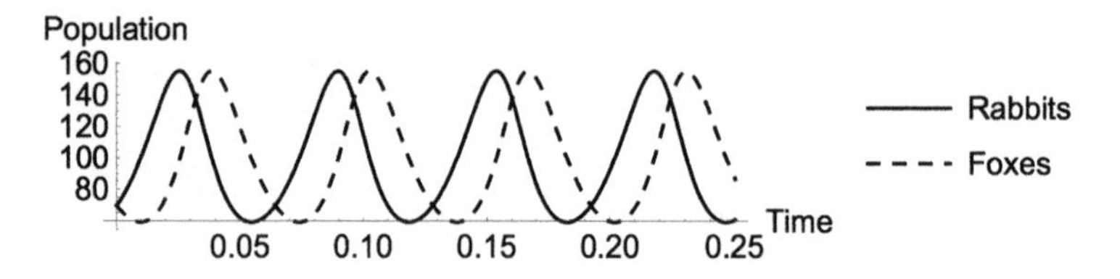

Figure 48: Oscillating populations of rabbits and foxes.

Initially, the rabbit population grows quickly because of the small number of foxes. Eventually, the surplus of rabbits causes the foxes to gorge themselves and the fox population rises quickly, too. Overcome by the large number of foxes, the rabbit population falls, and the fox population soon follows due to a lack of food. The cycle then repeats itself.

If the system starts out with different initial quantities of rabbits and foxes, the oscillation changes in size. To visualize how the system would behave at any given combination of rabbits and foxes, we can plot a grid of arrows that represents how the system will change from that point (see Figure 49). For example, the arrow at 10 rabbits and 200 foxes points downward, which means that the population of foxes will decrease and the population of rabbits will stay mostly constant.

In contrast, the arrow at the point corresponding to 200 rabbits and 200 foxes points up and to the left: the rabbit population will decrease and the fox population will grow. Each of the three gray curves shows a possible "path" that a system could take; any system that starts on one of those gray curves will continue circling counterclockwise around that same path forever.

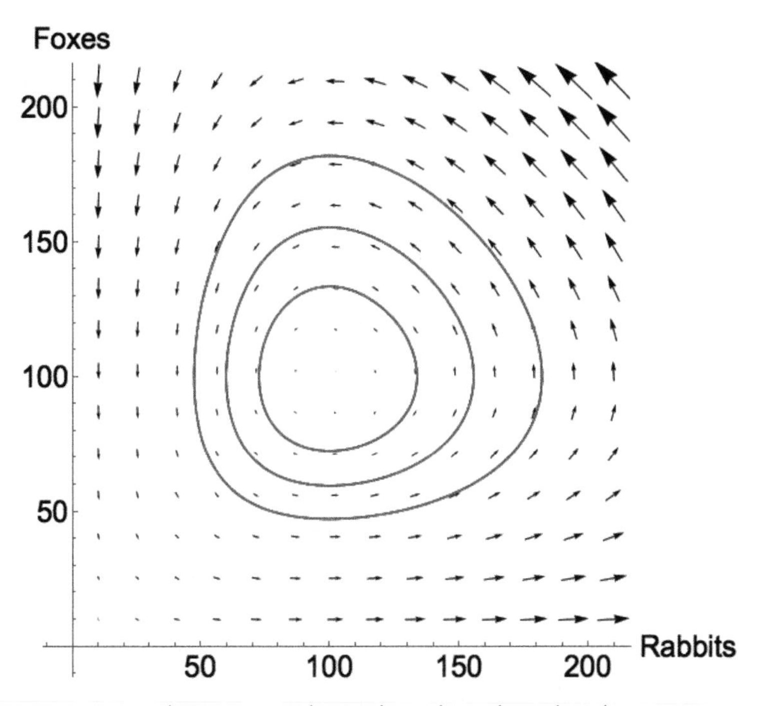

Figure 49: The system "orbits" along the indicated paths.

Of course, real ecosystems are much more complex than this simple model. Natural systems often involve hundreds of species that interact with one another. So what kind of behavior should we expect from large ecosystems?

Before the 1970s, it had long been assumed that ecosystems that consisted of many species with many interactions among them would tend to be stable against disturbances, such as the loss or introduction of some animals. But in 1972, an Australian scientist named Robert May published a widely discussed paper that challenged this assumption.[24] Indeed, May's paper suggested that a randomly assembled large ecosystem is actually virtually certain to be unstable.

To think about large systems in general, May used a simple mathematical abstraction. In May's model, the interaction of every pair of species is described by a number. A negative number means that more of the first species tends to cause the population of the second to fall, and a positive number means the first species promotes the growth of the second; a zero indicates no interaction. May considered an ecosystem of many species, all of which randomly interact with one another; that is, the numbers describing their interactions were chosen at random. He then imagined that all of the populations reach a point where they are balanced—where none of the species are growing or shrinking with time. (In math, this is called a steady state.) Then, May asked how the system would respond to a small disruption: if one of the species' population increases slightly, say, the system could either return to the steady state, or it could fall apart. A system that returns to the balance point after a slight perturbation is stable;[25] in contrast, an unstable system deviates wildly from its steady state with even very small perturbations. This is like gently pushing a person doing a yoga pose: will she regain her balance or fall over?

May proved mathematically that large systems where the parts are connected together at random are essentially guaranteed to be unstable;[26] they will almost invariably fall over when pushed. After a small disruption, the populations of some species may fluctuate wildly, or some may go extinct. This result doesn't mean that all large ecosystems must be fragile—a non-random system could still be stable—but the result challenged scientists to understand the contexts in which stability can arise. These ideas also have implications for the stability (or lack thereof) of human-constructed systems, such as the global financial system.[27]

Real ecosystems are not simply randomly connected; they have structure. Imagine an interaction between two species, again coded as a number. If species A eats species B, it might have an interaction of -1 because having more of A will reduce B. In a random system, the reciprocal effect of B on A is as likely to be -1 as it is to be 1, but that doesn't match what we see in nature. If A eats B, A is likely to get a positive boost from B's presence—which means that when the A-B interaction is negative, the B-A interaction is likely to be positive. More recent mathematical work suggests that networks with many of these predator-prey relationships may be more stable than their random counterparts[28]—and it may be that this structure saves natural ecosystems from instability.

And while the instabilities and oscillations scientists observe in natural populations could have various causes, the consequences of such oscillations can be serious. As an extreme example: every 48 years, the Indian state of Mizoram is utterly overrun with rats. When the rats come, they eat everything. In 1958, the rats wiped out the local crops and food was scarce. This caused a famine that killed nearly 15,000 people,[29] and the widespread hunger led to unrest: "The resultant famine provoked a 20-year guerrilla war between the disgruntled Mizo people and the federal government in New Delhi," reported the *London Daily Telegraph* in 2004 as Mizoram braced for another outbreak.[30] Historical records kept by the British colonial government seem to indicate that every time the rat flood arrived, the results were both epic and devastating.

Scientists can trace the cause of this ratpocalypse to the rats' interactions with a bamboo species native to the area, which flowers and bears fruit every 48 years. In most years when the bamboo doesn't flower, the number of rats stays fairly small. They scrape by, eating whatever they can, but there's no big source of food that would allow them to grow out of control. But in those years when the bamboo does flower and produce fruits, the rats gorge themselves on the bamboo fruit and breed furiously, swelling their numbers enormously . . . until the bamboo fruit runs out.

Once the rats have eaten most of the bamboo fruit, the population is far larger than the normal food supply can support, so the rats raid the local farmers' crops. That huge mass of rats utterly demolishes the farmers' fields. It is perhaps for this reason that, according to newspaper reports, the name for the bamboo, *mautam*, is synonymous with "famine" for some locals.[31]

If the rat flood comes every 48 years because those are the only times they can gorge themselves on bamboo flowers, that suggests an obvious question: why does the bamboo keep this odd schedule in the first place? It turns out that the bamboo might, in turn, flower every 48 years because of the rats.

It seems that the rats and the bamboo may be connected together in a system that has evolved towards cyclical behavior—a clock. Think of it like this: from the bamboo's perspective, its "goal" is to have as much of its seed as possible survive to become the next generation—or more accurately, a bamboo plant that has more of its seeds survive will eventually beat out less successful bamboo plants through the process of evolution. The rats are one of the big obstacles to this survival, however, due to their appetite for bamboo seeds. In this war between the bamboo and the rats, perhaps the bamboo developed a strategy: shock and awe. Instead of releasing seed every year and giving the rats a reliable source of food, they would release it only occasionally. And when they did release seeds, it would be intense. With all of the bamboo plants releasing their seeds all at once, the rats would be overwhelmed and totally sated. While a lot of seeds would be eaten, a lot would also survive by getting lost in the shuffle.[32]

Once this strategy develops, it is probably self-reinforcing. Imagine a situation in which most of the bamboo plants bloom every 48 years, but a few rogue plants flower at some other time, perhaps yearly. The bamboo plants that flower as a group have at least some of their seeds survive the rat menace because the rats are overwhelmed by the amount of fruit available. The few plants that flower out of sync are in trouble: because the rats have less food from other sources, much of the rogue bamboo's fruit is eaten by the rats. In this way, the rats might punish any bamboo that flowers out of sync. This could provide the selective pressure necessary to maintain such an incredible clock.

PART III
APPLICATIONS

CHAPTER NINE

Build Me a Buttercup

Using Synthetic Biology to Make Diesel Fuel, Programmable Cells, and Malaria Medicine

Somewhere in Dr. John Love's lab at the University of Exeter in southwest England are tiny plastic plates filled with bacteria growing in less than a thimbleful of liquid. And even though that liquid is mostly just water, salt, and sugar, the bacteria are doing something incredible: they're making tiny amounts of diesel fuel.[1]

"I've always really been quite interested in primitive plants," Love says, even since his undergraduate years. He explains his fascination as mostly academic, but doesn't rule out other influences. "I suppose coming from Scotland, there's a lot of moss around," he jokes—hence the focus on plants. But when it came time to set up his own lab, conversations with friends in the oil industry convinced him that some of the plants he had studied might be used to improve our fuels. With a new focus on making fuel, Love set out to explore what biology could offer.

He started by studying organisms that naturally produce oily, fuel-like substances, but he eventually decided that was the wrong strategy. The problem was that the fuels they were making weren't very compatible with today's engines. "The general biodiesels that we make are oxygenated," he explains, "and that is very corrosive." Simply burning corrosive fuels could damage existing engines. Eventually, he concluded that making them usable was just too expensive. "Regardless of the level of production that these organisms were capable of, the cost of processing the oils afterwards was very high," he recalls, and that pretty much made these types of biofuel a nonstarter for commercialization. Love needed a different approach.

Then came the realization: if he could make a fuel that didn't need a lot of post-processing—one that could be used directly in modern combustion engines—that fuel might stand a chance. Love knew that bacteria naturally produce molecules called free fatty acids, and these greasy compounds are chemically quite similar to the molecules that make up diesel. If it just had the right enzymes, a bacterium might be able to make diesel fuel from these free fatty acids. And luckily, he knew just where to find those enzymes.

"We had been working for ten years previously on characterizing every single oleaginous organism," he says. (Oleaginous organisms produce fatty, fuel-like molecules.) This work made him pretty familiar with the enzymes that nature has already come up with to work with diesel-like molecules. Out of this vast toolbox, he identified the enzymes he needed, and he and his team went to work. They took the genes for these enzymes and put them into regular bacteria. Putting a gene into a bacterium can be as simple as shocking it with electricity while the gene floats around in its environment, but the next part was much more difficult: Could these enzymes really enable the bacteria to produce the major chemicals that make up diesel fuel?

After lots of work, it was finally the moment of truth. "We thought this *could* work," he explains. "But it was getting that GC [gas chromatography] trace back," he recalls, referring to the instrument they used to test for the presence of the diesel fuel compounds, "and you see the little blips for the alkanes, and you think, 'Bloody hell, it *has* worked!' And then you do it again, and yeah, it's worked again! And you get somebody else to do it, and it works for them too! That was awesome." You can hear the energy in his voice, even now. "We weren't really sleeping very well that week."

The concept worked, at least at a very small scale, and since then, Love and his team have been trying to optimize the system to produce more than just a tiny amount of diesel. There's a huge leap to be made between making small amounts of fuel in a lab and producing it on the scale required for industrial or consumer use, as Love himself is quick to point out. "The difficulty, if you like, in all of these processes is scaling up," he notes. "Every time you increase it by an order of magnitude in scale, problems that weren't problems previously suddenly raise their ugly heads." They've run into issues of heat dissipation, nutrient availability, and oxygenation, and there most likely will be future problems no one can anticipate. Love's bacteria will have to increase in efficiency by at least a few hundred-fold before they become even remotely feasible to use commercially, and it remains to be seen if that's possible.

Even if Love and his team can produce enough diesel with this method, it's not clear that this biofuel will be economically viable. Right now, Love's bacteria feed on sugar, and at the time he and I spoke, sugar was about five times as expensive as diesel. "You're losing 80% of your value, and you're actually doing a very expensive process to do that," he complains. "You're taking dollars and changing them into cents." Ultimately, though, it should be possible to lower or eliminate the cost of the raw materials by getting the bacteria to engage in what Love calls "extreme recycling:" growing on waste products like sewage, compost, or other scraps. "You know, poo," he says and winks. Diesel from poo.

Despite the challenges ahead, Love seems to remain both realistic and optimistic. "I don't pretend that this will replace petrol," he says. "What it will do is it will buy us time, and it will make the supplies that we have more sustainable." He reiterates: "I'm not trying to solve the world fuel crisis." But this technology, if it were to succeed, certainly wouldn't hurt.

John Love is one of a new breed of scientists called synthetic biologists who are approaching biological systems in a whole different way: as machines that we can build. Synthetic biologists are systems biologists who aren't content just observing existing systems—they want to make new ones. That's an incredibly exciting way to think about systems: as Lego sets.

The past century of biology research has produced a vast catalog of systems and parts that occur in nature. They don't fit together quite as easily as Lego blocks, but we are starting to realize that we can use them to construct biological systems that have the potential to produce medicines, sense environmental toxins, and improve manufacturing processes. Building new biological systems, hence the term "synthetic biology," is popular right now in part because its premise is compelling. What if we could indeed design biological systems in the same way we design electrical circuits, or cars, or factories, or toasters?

In some sense, building anything can show that you know how the end product works . . . or that you don't. Building things from scratch has a way of revealing faulty assumptions and missing pieces, so biologists often try to reconstitute a system that they're studying in a test tube to show that they're not missing any steps of the process.

For his part, Love's work represents some small example of what biology might eventually make possible. His team and dozens of other researchers had spent many years working on organisms that had useful components for doing the kinds of chemical reactions Love needed to make happen. So when the time came to actually make the diesel-producing bacteria, he was able to choose the best parts from a catalog of possibilities. "We're now able to use natural parts to design things that don't exist in nature as an entity," he says. "Through synthetic biology, we're no longer tied to the limitations that nature gives us."

Dr. Michael Elowitz was one of the scientists who helped convince the world that building biology was something worth considering. In 2000, Elowitz was growing bacteria that he had modified to do something special. The cells looked perfectly normal, except that some of them would glow green under a blue light. Glowing cells themselves are not unusual; scientists often use proteins that glow under blue or green light to tag parts of a cell so they are easy to observe. But if you watched Elowitz's cells for a long time, something more interesting would happen. Over the course of three hours or so, the cells would glow, fade, and glow again—over and over in a slow pulsing rhythm!

Elowitz is now a professor of biology, biological engineering, and applied physics at the California Institute of Technology. Around the turn of the millennium, when he was growing those glowing cells as a postdoctoral fellow at Princeton University, he was frustrated. To him, even some successful research studies felt like they were missing something. "After doing these biology experiments, you produce some kind of model, or picture of what you think is going on," he says. That model is an attempt to explain how the thing under study actually works, but it's almost impossible to show conclusively that the model is right. We can predict how the system will respond, "but it's really hard to say whether that set of interactions is actually sufficient to produce the kind of behavior that people have observed," Elowitz says. We might think we know how something could work, but the only way to really find out is to build it.

Elowitz wanted to design a set of genes that would perform a desired function when put together in an organism. After some deliberation, he and his adviser, Stanislas Leibler, settled on building a simple system that would be feasible to construct[2]: a regular oscillator, which you can think of as a kind of clock—remember, a clock is really just some process that repeats, or oscillates, at a regular interval. So Elowitz set about to make an oscillator inside of his cells that glowed green at regular intervals.

Out of all of the possible systems, why build a clock? First, it was closely related mathematically to problems Elowitz was used to working on in physics. And second, it seemed like an interesting system because rather than simply responding to environmental inputs, a clock is constantly doing something visible. "It spontaneously generates its own dynamic behaviors from the inside," Elowitz says, and that made this kind of device exciting to observe. Plus, he thought it would be fun.

Elowitz started planning. "It took me a little while to think about different designs that might do this," he said, but eventually he settled on a plan. Elowitz would hook up the genes for three repressors, or proteins that turn off the expression of some other gene, in a circle. Elowitz sometimes compares it to a game of Rock, Paper, Scissors: gene Rock makes a protein that turns off gene Scissors, which makes a protein that turns off gene Paper, which in turn turns off gene Rock. Simple mathematical equations predicted that this system should oscillate under certain conditions. They called this device the repressilator, from "repression" and "oscillator," and they built it by putting those genes Rock, Paper, and Scissors into a bacterial cell.

And it worked.[3] It wasn't a great clock: the time between ticks often varied by 40 minutes or so from the average of 160 minutes. But it was a clock nonetheless, and it had gone from an idea to a design to a working system in a cell.

Elowitz's glowing clock was a proof of concept, and it was pretty successful in that regard. The repressilator became a well-known early example of biologists building circuits in cells rather than just studying them. And while later work by many different scientists has refined the initial design and made the clock work better, this is commonly recognized as a turning point for synthetic biology: the prototype showed that biology was indeed something we could engineer.

The repressilator, while relatively simple, took a lot of work to design. While spending months to conjure up a simple three-component system might be acceptable for a proof of concept, the time it would take to build very complicated circuits could be prohibitive, or even impossible. That's why some scientists are looking to computers as a model for constructing biological systems.

Computers are incredibly powerful machines, but underneath all the fancy windows and graphical interfaces, your computer really only understands a small number of simple operations. Fundamentally, it uses a binary code: every program is ultimately a long string of ones and zeros that tells the computer what to do.

In order to make the computer do something new—that is, to write a new program—you need to give it a series of ones and zeros that spell out the very simplistic steps of what to do: read a number from memory at location 1, read another number from location 2, add those numbers together, and write the result to memory at location 3.

Meanwhile, humans like to think on another level. When programming a computer, for example, you would like to be able to write something like *for every number from 1 to 10, show that number on the screen*, but the computer needs to get something like *10010010 11111010 10111010 11000000 10101010 00010111 10101010 01001011 11011111 10111010 10111010 11111110 00101010 10001010 00011010 10001000 01001010 10111010*. Writing large programs in that binary machine code is basically impossible, so computer programmers use a tool called a compiler.

The compiler is both a translator and an expander. It takes your simple, high-level description—"add one and two, then store the result in memory"—and it does two things to it. First, the compiler breaks down your program into simple instructions that the computer can understand, and then it translates those simple instructions into machine code. The ability to design software at a high level has led to the computer revolution—and a similar revolution would be really useful in biological machines.

Bacteria and other organisms don't speak in ones and zeros, but they have another low-level language that is hard for humans to think in: the language of DNA. As much as it is tedious and time-consuming to write computer instructions in machine code, it's arguably much worse to write in the genetic code. While we fully understand how computer chips work—after all, we designed them—our grasp on how gene expression is regulated is still a work in progress.

One early attempt at helping humans "speak DNA" was Biocompiler,[4] the brainchild of Dr. Jacob Beal and Dr. Ron

Weiss at the Massachusetts Institute of Technology. Their idea, shared by some other scientists, was that one day you will be able to write a program that is then automatically translated into the genetic code necessary to produce the system you designed. For example, the code snippet they used to illustrate their idea, *(green (and (not (aTc)) (not (IPTG))))*, means the cell should glow green if it does not detect the chemicals aTc and IPTG. More recent attempts have moved this idea closer to being a practical tool by automating more of the design process,[5] but this technology is still in its infancy.

Another take on this concept that looks promising is systems that use DNA itself to do computations. For example, researchers at Caltech, the University of Washington, Harvard, and the University of California, San Francisco, are part of a collaboration called the Molecular Programming Project. They use short strands of DNA that can interact with one another to do simple logic, much like computers do at the most basic level. Their "programs" are sets of DNA strands: strand A will unfold only if B and C are around, and when A opens it can then open D and E, and so on. The team has also developed computer programs that anyone can use to design these types of DNA logic circuits.

If biological programming ever takes off, it could potentially make solving many other biological problems easier: it would be a framework and a set of tools that could pave the way for other work yet to come.

War. What Is It Good For? Apparently, Antimalarials.

More than 300 million people worldwide have malaria. The disease, spread by mosquitoes, claims more than a million lives every year,[6] and malaria helps keep those it spares trapped in poverty.[7] As Bill Gates has noted, the scourge of malaria means that mosquitoes kill more humans than any other animal, including other humans.

It is so bad that many scientists think that malaria is probably responsible for keeping sickle cell disease in the gene pool.

Sickle cell disease is a heritable condition in which the red blood cells take on an altered shape that resembles a sickle. The disease arises when a person has two variant copies of a gene that encodes a component of hemoglobin. In contrast, having one variant copy and one normal copy of this gene changes the shape of the red blood cells in a way that does not usually affect the health of the person but does provide some immunity to malaria. The malaria parasite attacks red blood cells directly, and this milder change to those red blood cells disrupts the malaria without endangering the person. It is thought that sickle cell disease hasn't been purged by evolution because this "milder form" gives immunity to malaria, even though sickle cell disease itself can lead to crippling lethargy and early death.[8]

We do have drugs that can combat malaria, even though their expense sometimes keeps them out of the hands of those who need them most. And strangely enough, one of our weapons in the war against malaria came from the Vietnam War.

The militaries on both sides of the conflict in Vietnam were very interested in developing medicines to help deal with the problem of drug-resistant malaria. The Americans, under the leadership of the Department of Defense's Walter Reed Army Institute of Research, developed a drug called mefloquine. The North Vietnamese lacked the resources to undertake a huge drug development effort, so they turned to Chairman Mao Zedong of China for help.[9]

In an effort to aid their allies, the Chinese launched "Project 523" to find new antimalarial drugs. Part of Project 523 involved screening plants that were used in traditional Chinese medicine for possible leads. A

Chinese scientist named Tu Youyou and her colleagues investigated more than 2,000 possible remedies before settling on a compound extracted from a native Asian plant, sweet wormwood. This compound, artemisinin, turned out to be a highly effective antimalarial drug. Tu Youyou received the 2015 Nobel Prize in Physiology or Medicine for this work.

And in a weird twist on the history of these compounds, artemisinin, the Chinese drug, and mefloquine, the American compound, work better when used together.[10] Since there is a danger that the parasite will develop resistance to artemisinin if the drug is taken alone, it's sometimes given with mefloquine to help prevent the malaria parasite from developing resistance to the drugs. So while Communist China and the United States didn't always get along, their pharmaceutical drugs did.

Artemisinin has turned out to be an important tool in battling malaria today, but it's relatively expensive to produce and relies on sufficient supplies of sweet wormwood. Synthetic biology may turn out to be of some assistance here. Jay Keasling, a professor of chemical engineering and bioengineering at the University of California, Berkeley, and his colleagues thought that there might be a better way of making artemisinin than extracting it from plants. Instead, they programmed yeast to do the work for them.[11]

Yeast naturally make a precursor chemical that is distantly related to artemisinin, so Keasling's team started there. They genetically altered the yeast to beef up the enzymes that make this precursor, and they reduced the levels of a different enzyme that uses this precursor to produce other end products. Next, they copied the genes from the sweet wormwood plant that produce artemisinin naturally, and they put those genes into the yeast that they had modified to produce high levels of the precursor. And lo and behold, the yeast started making the drug.[12]

Producing artemisinin in a lab was a long way from making sure it was safe and efficient to use as medicine, but in 2014, after a decade of work by Keasling's team and others, the pharmaceutical company Sanofi began shipping artemisinin produced using these synthetic techniques. This success demonstrates the promise of using biology as a factory: pharmaceuticals and other chemical products that are limited only by our ability to engineer them.

Until programming cells becomes a reality, synthetic biology will require a lot of work—but the rewards are enormous. One promising possible direction is the creation of biosensing organisms that sound the alarm when they detect environmental toxins. And in some ways, this idea borrows from coal miners.

Mining coal is dangerous. Workers face mine collapses, black lung, or fire, and one threat that's particularly pernicious because it's invisible: carbon monoxide. Since it's colorless and odorless, carbon monoxide can kill without much warning. Today's coal miners have electronic carbon monoxide detectors, but the precursor was a biological sensor: a canary.

The idea of the canary in the coal mine seems to have come from John Scott Haldane, who in 1913 proposed using small warm-blooded animals to test for danger.[13] Canaries were preferred because their signs of distress were quite visible.[14] A canary "would sway noticeably on his perch before falling," a signal of deadly import.

Using an animal to sense danger is a clever idea, but it's of limited utility if the danger in question doesn't produce visible signs of distress in the way carbon monoxide does. There are a lot of toxic or dangerous substances that take years to kill an organism, but we would like a biological sensor to be able to detect them. Unfortunately, it's not really practical to wait a few years to see if the canary gets cancer or not. So what if we could get other kinds of organisms to tell us when something's wrong?

One potential solution might come from a modified version of the bacterium *E. coli*. Scientists have devised synthetic systems within these bacteria that can monitor a number of specific air, water, and soil contaminants. Besides their potential use as a detector of specific toxic chemicals, though, we might also be able to use them to screen all of the chemicals the cell comes into contact with for properties that might pose a threat to human health, such as the propensity to damage DNA. We care a lot about DNA damage because many things that cause DNA damage are carcinogens. They can mess up the systems that control cell division and growth by causing mutations in the genes that control those processes. This can disrupt the cell and cause it to divide out of control, which is the hallmark of cancer. A bacterium that could report back when it encounters something that damages its DNA could therefore alert us to be careful of that environment or chemical because of the potential cancer danger.

The question is how to figure out when the bacteria are in trouble. Fortunately, there's a simple system that already does most of the work for us. When bacteria—and essentially all other cells—detect that their DNA has sustained some damage, they mount a concerted response to try to repair the damage. Scientists can then listen in on that DNA damage response, part of which is colorfully called the "SOS response" in bacteria. To do so, they use proteins that are somewhat like the glowing proteins Michael Elowitz used to watch the progress of his clock, but these proteins produce light themselves instead of just fluorescing under illumination. Researchers can, for example, take the regulatory elements that normally produce a DNA damage response gene, called *recA*, and hook it up to a set of genes that produce light. Scientists have tried this,[15] and indeed, these engineered cells are able to glow when they start getting into trouble.

Generally speaking, there are a few possible advantages to building a biological system for this task. Since DNA damage is the thing we want to measure, it seems natural to use a living system as the measuring device.[16] And though this is about the simplest possible engineered system you could build to detect some environmental toxin, it should be relatively straightforward to design more complex systems that respond to the presence of multiple toxins or to chemicals with multiple suspicious characteristics. Bacteria also have the advantage that they don't break in the way that classical instrumentation does, and they're very cheap to maintain. They can survive in a wide range of environments, including in the presence of really nasty toxins. They're pretty ideal for going to precisely the places that humans don't want to go and sensing what those environments are like—the ideal environmental alarm system.

Other groups are attempting to cut the bacteria out of engineering altogether and focus on the DNA. In fact, if you walked in during the middle of Peng Yin's lectures at the Wyss Institute for Biologically Inspired Engineering at Harvard University, you might momentarily think that synthetic biologists are really excited about fonts. A professor in the Department of Systems Biology at Harvard, Yin often shows a slide that, at first glance, just looks like a bunch of symbols. It's as if you lined up every character available in the Times New Roman font and threw them on a slide. All of the letters of the alphabet are represented, and there are cubes, pyramids, and even a tiny smiley face. But these are actually pictures taken with a high-powered microscope—and those tiny letters, pyramids, and smiley faces are made entirely out of DNA.

We think mostly about the information content of DNA—the genes and regulatory sequences that it encodes—so it's easy to forget that it has some convenient structural properties, too. Most usefully, DNA sticks to other DNA if the two molecules have complementary sequences. As mentioned previously, DNA is made of small bits that we abbreviate as G, C, T, and

A, strung together in a long line. G's stick to C's and A's stick to T's, so the sequence GCAT will tend to match up with the sequence CGTA, for example. Because of this property, you can simply throw a bunch of DNA sequences into a tube and, under the right conditions, they will match up those complementary parts and form a structure. In fact, evolution already makes use of this ability to make structures in DNA's closely related cousin, RNA. Among many other functions, cells use RNA to make the factories that produce proteins, for example—those same protein factories that use delays to make proteins so accurately, as discussed in Chapter 3. It is RNA's ability to stick to other, complementary sequences of RNA that makes all of this work.

So what exactly does one do in order to build something with DNA? For simple structures, all we have to do is design the DNA sequences so that they can fit together in only one way: the structure we want. And since scientists have spent many decades developing tools to work with DNA, there are a lot of existing methods for making any given DNA sequence, as long as it's not too long. Thanks to these tools, it's relatively easy to execute our designs, even if they are quite complicated. In Yin's lab, researchers design hundreds of strands of DNA that fit together in only one certain way; then they make those sequences, throw them together in a tube, and the DNA self-assembles into the shape.

The point of these kinds of projects is to prove the principle that we can manipulate DNA to make it do what we want. DNA structures can already be used to make intricate designs and even tiny machines, like bipedal DNA "walkers" that could be used to shuttle tiny components around. Other DNA structures could be used to package drugs, do nanofabrication, or make any number of other tiny molecular machines. But the main takeaway is that there are many biological materials that we might wish to make novel systems out of, whether it's proteins that produce cellular clocks or the DNA itself.

Race Day

Some people race horses. Daniel Irimia, an assistant professor of surgery at Harvard Medical School, races slime molds. Slime molds are tiny, single-celled organisms that do all sorts of weird things, like group together and form stalks in order to reproduce: they look like tiny cattails sticking up from the forest floor. These creatures, more accurately referred to as "social amoebas," aren't the most conventional choice of creature for racing, perhaps, but Irimia has a reason for choosing them: tinkering with these cells may help us eventually better understand arthritis, severe burns, heart attacks, and even cancer.

Irimia's social amoeba of choice to study is *Dictyostelium discoideum*, used in labs the world over and often lovingly called "Dicty" by the scientists who study it. Dicty is a useful model for studying how cells migrate in response to chemical signals from their environment. This migration is really important in human cells, such as a type of white blood cell called neutrophils. Irimia explains that neutrophil mislocalization can lead to some really bad outcomes: "That's when you have neutrophils going into a tissue and releasing their enzymes like they were attacking a bacteria, and destroying normal tissue instead." And as a general rule, we'd like our bodies not to destroy normal tissue accidentally.

Studying the systems that govern cell motility to try to get some understanding of these processes is challenging, Irimia says, because there are many molecules involved and motility is governed by a complicated system. Indeed, there are dozens of labs around the world that study how cells move. Unfortunately, according to Irimia, everyone seems to tinker with different parts of the system or have different ideas about how to think about cell motility. To get everyone on the same page and spur a little friendly competition, he decided to hold a race.

He challenged twenty labs from around the world to submit Dicty cells that they had genetically engineered or otherwise modified to move more quickly and efficiently through a tiny maze toward a chemical signal. The goal, he says, was to "bring all the work that was being done in the field of cell migration together," and this was a fun way to do it. All of the field's research could be evaluated on the same maze, the thinking went, so it would be easy to see what works and what doesn't. Irimia hoped that his race would help show how we can make cells like Dicty faster and smarter.

So on a Friday in May 2014, "all hell breaks loose," as Irimia put it, laughing. Cells came streaming in, packaged in small tubes in FedEx envelopes. Each team's cells got a shot at the maze, and the results were broadcast on the race's website. Ultimately, cells raised by Arjan Kortholt and Peter van Haastert, from the University of Groningen in the Netherlands, emerged victorious. For their efforts, they took home the $5,000 grand prize and eternal glory.

The champions also proved an old adage: slow and steady did win this race. The most successful racers were the cells that could follow the chemical signal through the maze most efficiently, not the ones that moved the fastest. The winning cells were genetically engineered to make more of a protein that helps them move in the correct direction when the chemical signal they're tracking is weak.

A race was an entertaining way to raise awareness of problems in cell motility, but more generally, Irimia hopes the Dicty World Race will encourage people to tinker with biology. There has even been interest from nonscientists, he says. "When you do science, usually you are alone in the lab and no one cares what you're doing," but this race changed that, at least for a moment. "It's beyond what I imagined it was going to be."

The twin spirits of curiosity and competition have propelled scientists to make all kinds of novel systems by changing or adding to an organism's genetic code. Some scientists have gone further still: instead of simply rewriting the existing genetic code and making new machines from known parts, Floyd Romesberg, a professor of chemistry at the Scripps Research Institute in California, and his colleagues added to the code itself.

Romesberg is fascinated by the fact that our DNA contains the information we need to build an organism. "I just can't escape this idea that everything in life that we see today—all the diversity that we see—is encoded in a four letter genetic alphabet," he marvels.

Think about it this way: this book is written in a code of sorts. English has 26 letters, plus punctuation and spaces, and these pages convey information by using those letters to form words. And much like a traditional alphabet, DNA's A, T, G, and C bases together form a language, the genetic code, that carries most of the instructions for making a new copy of that organism and making it work. Inside genes that encode proteins, for example, each set of three letters is a "word"—an instruction to add a certain type of protein building block to the end of the growing protein chain.

This way of writing information has been fantastically successful; it is responsible for all of the diversity of life that we see today. Every blink, sprint, flex, and yawn that has ever been simply couldn't have happened without the information carried in DNA. It is safe to say that the four bases of DNA are the most successful language ever invented. But despite their success, four bases just weren't enough for Floyd Romesberg. He wanted six.

Expanding the genetic code could allow us to expand the parts list available to synthetic organisms. By modifying the genetic language, we could allow for the inclusion of some components that might not be useful or practical for living things to use, but might be of use to us. By expanding the list of possible words, we could reserve some of those words for totally novel protein building blocks.

Romesberg is especially interested in the possibility of making proteins with different protein building blocks for use in medical treatments. Using proteins to treat diseases has become much more popular recently among pharmaceutical drug companies, in part because they're easy to make and we can actually use evolution in the lab to help design them: scientists make huge numbers of random modifications to the protein, then select those that perform best, perhaps by binding tightly to a target. Repeated cycles of this process can evolve proteins that bind to the target much better than the original protein did.

Proteins aren't perfect, however: "A downside of protein therapeutics is that you're stuck with what the protein has, and proteins are built, more or less, from 20 amino acids," Romesberg says. But if you had extra words, you could hypothetically add pretty much anything into the protein chain. "Could you incorporate an unnatural amino acid whose side chain is something like, I don't know, penicillin? Or some small reactive center that you hope is going to impart the protein with the potential for whole new sets of behaviors?" The possibilities are intriguing. We don't know exactly how we might best use such a technology yet, but it's pretty clear that there would be lots of applications.

But before we can do any of that, we have to make it work. The idea of using unnatural base pairs has been around for a while, but it is hard to do for a lot of reasons. Billions of years of evolution have refined the machinery that reads and replicates the standard four bases, so making an organism that uses extra bases requires making tweaks to a very-well-tuned system. It means making the existing DNA-copying machinery play nicely with the new letters. It means making sure that the aggressive mechanisms that the cell has developed to protect its DNA don't see these bases as an error in the DNA and promptly excise them. It also means getting those bases into the cell—or, eventually, getting the cell to make them for itself. All of these are substantial challenges—or at least, they were before Romesberg and his team solved them.

Romesberg and his team started with two new bases, which they call X and Y.[17] In order to get an organism that can use these bases as part of its genetic code, they had to (1) get the X and Y bases into the cell, (2) get the cellular machinery to copy DNA with these X and Y bases in it, and (3) keep the cell from seeing the X and Y bases as errors and eliminating them.

The group set to work on the first challenge: transporting the X and Y bases into the bacterial cells. After a lot of work, they found proteins from certain kinds of algae that they thought might do the job. They screened through many of them, and eventually one worked; when Romesberg's team put the protein into their bacteria, the X and Y bases were able to make their way into the cells. They had cleared the first hurdle.

Expecting a similar slog for the next two challenges, Romesberg's group ran some experiments to see how bad the problems were . . . but there were no problems. Once they solved the first challenge, everything else just sort of worked. "It's kind of an anticlimactic story," he admits. "Here we were thinking, okay, we've got our first step done, we've got two more to do," he says, "and then a week or two later, we're like, crap, this works." Indeed, those bacteria were able to copy a bit of DNA that included the new bases and pass it stably along down the generations. After 15 years of hard work optimizing X and Y in test tubes, it turned out that getting a cell to use these new bases was relatively easy.

These were the first organisms in the history of our planet to propagate information through a partially artificial genetic code. They represent our first halting steps into a world where natural systems and systems of our own design commingle to build mini-factories, sensors, biological computers, medicines, and things we haven't even thought of yet. And there are a few scientists who want to test our knowledge by going even one step further: they want to build biology from the bottom up.

If we can supplement the genetic code and build simple circuits, it is not too much of a stretch to go the rest of the way to full manipulation of an organism's genetic code. If we were to replace all of the DNA in, say, a simple bacterium, we would create a new type of bacteria with different genetic instructions. In order to do that, we would need to build a genome from scratch—and as it turns out, we can do that now.

J. Craig Venter rose to prominence in the public eye as the head of a private competitor to the government-funded Human Genome Project known as Celera Genomics. After the genome was completed, Venter turned to other projects, such as sailing his yacht around the world and sequencing the bacteria he found in the sea along the way. Venter is a prominent and occasionally polarizing figure, and he's put a lot of money and effort toward scientific projects. One of those projects was synthesizing a whole genome.

Scientists have been able to make short bits of DNA for a long time now, and there are a few decently reliable methods for stitching small bits into bigger bits. Before 2010, no one had taken these technologies and pushed them to their limits to make a really massive sequence, though. Venter and his team decided to do this to build the full complement of DNA needed to run a single cell.[18] The genome was small—only about a million bases, compared to the three billion bases humans have—but it was created entirely from scratch, starting with just the sequence information. As a template for their new genome, the team used a small bacterium similar to the one whose genome they were replacing. They made a few changes to give this synthetic genome its own "fingerprint," but mostly stuck with the template sequence. After lots of effort, though, Venter's team had built a synthetic genome.

Once the full sequence was assembled, it was time to test it out. Venter's team took an existing bacterium, removed its genome, and inserted the newly created genome in its place. It worked. The bacteria with the replacement genome grew and divided just like one would expect if it had contained the synthetic genome all along. Effectively, Venter's team had changed one type of bacteria into another similar type by replacing its entire genetic code.

It's worth noting that Venter's project was a bit different from making bacteria that can produce diesel fuel, as we talked about at the beginning of this chapter. In that case, John Love took an existing organism and added a new system on top of it, like retrofitting a car with an aftermarket stereo that you've designed yourself. Venter and his team took the blueprints for an existing car, then built the engine from scratch and dropped it into an existing chassis—and the car still ran. These are different flavors of synthetic biology, but both could one day prove useful in changing how we make medicines, how we do science, and how we think about the world.

People sometimes wonder if types of synthetic biology like Venter's constitute artificial life. That depends what you mean by "artificial life," but probably not. Venter built a genome from scratch, but it was basically entirely the same sequence as a natural genome—this was a feat of genome construction, not design. Furthermore, this work still relied on the cellular machinery of a cell that had been stripped of its own DNA to incorporate the artificial genome. The researchers needed the donor cell in order to allow the new genome to run a cell and reproduce for the first time; his team built a long sequence of DNA, not a whole cell. Then again, every cell after the first one is made using the synthetic genetic material: whenever the cell divides, it has to use the information from its new genome to make the parts for a new cell.

Venter's team inserted a genome that they had built mostly based on existing sequences found in nature, but in principle there's no reason that had to be the case. From there, it is a matter of incremental progress: change the blueprint a little bit, then a little bit more, and soon enough they could create something that is entirely unlike any life that has ever existed before. The big challenges of creating a truly new organism are that any genome that's too different from the donor cell's machinery might just not work and that designing a novel genome sequence from scratch is far beyond our current capabilities. But ultimately, despite the challenges, there's no obviously insurmountable theoretical obstacle to making organisms that are quite different from things that occur naturally. There probably won't be a single moment when "artificial life," whatever that means, is suddenly and indisputably created. Instead, our synthetic biological systems will become more and more different from existing life along a continuum.

And as we begin to understand the principles and parts that govern how biological systems work, building entirely new systems has become practical—and scientists and engineers have jumped on board with all sorts of exciting potential applications. If you ask John Love, that's no accident. "You know it's a very sort of human endeavor to create things, to build things, and I think that's what we're able to do now using natural parts," he says. Eventually, these types of techniques could change biology from a mystery to a tool.

CHAPTER TEN

More Than Just 86 Billion Neurons

The Science of the Brain, and How Connections among Neurons Make It Work

One of the best ways to appreciate the brain—and how little we know about it—is to watch a video of a patient who has a device implanted in his head. He's explaining the machinery that lets him turn off his Parkinson's disease with a switch.[1] "The probe that is placed into the brain," he begins, "comes level with the top of the nose and the ear." He gestures to the side of his head, and he traces a line down to his breast pocket. "And it is connected to a battery in my chest." He taps his finger to indicate an area that looks to be near his heart.

The man looks professorial, and he talks directly to the camera. He wears a bow tie, and from his calm tone, you would think that he was talking about a particularly nondescript book that he's just read. Instead, he is describing a piece of technology that drastically reduces or eliminates the tremors that Parkinson's disease causes.

Even though he has a crippling neurological condition, you wouldn't know it from looking at him until he says, "And I'll turn myself off now." There's a beep, and a moment when nothing happens. Then, his arm begins to shake uncontrollably. He tries to speak, but you can't hear him over the sound of his hand involuntarily smashing against his microphone. In the course of five seconds he's gone from *seemingly healthy* to *totally unable to function*. But he finally switches the device back on, and as suddenly as it began, his tremor is gone. He's back to being entirely normal—except for the electrodes in his brain, that is.

This patient is using a treatment called deep brain stimulation. Those electrodes that are implanted into his brain directly apply electrical pulses at an adjustable frequency to specific regions of his brain, and this stimulation can drastically improve the symptoms of Parkinson's. While it does nothing to stop the progression of the disease, it does get rid of the symptoms.

No one is sure exactly how deep brain stimulation works. We know that we can treat Parkinson's disease by putting electrodes in one specific place in the brain, and obsessive compulsive disorder or essential tremor by putting them in other places, but the mechanistic reasons why this helps the patient remain elusive. Like much of the inner workings of our brains, deep brain stimulation is still a mystery.

Fortunately, recent technical and conceptual advances have allowed scientists to start to understand what's going on behind the enormous complexity of the brain. But perhaps more so than any other biological topic, the brain needs to be understood as a system. The properties of the brain that people are most familiar with—thinking, consciousness, creativity, introspection, emotion—are totally inscrutable if you understand only one or two parts of the system. You can describe in perfect detail when and how a neuron "fires" an electrical pulse, or "turns on." But no matter how well you understand a single neuron, it will never tell you how consciousness works. Instead, all of the fantastic properties of the brain emerge as the result of the whole system somehow working together.

And while the brain is an unbelievably complex system, the reward for understanding it is immense. Besides the potential for developing new treatments for conditions such as Alzheimer's disease or schizophrenia, a full understanding of the brain would probably explain consciousness, memory, intelligence, creativity—just about everything you do with your brain would be fair game.

Insofar as most people have an idea of how the brain works, they think of it like a computer, but this is a rather misleading analogy. In your laptop, for example, there are about a billion electric switches called transistors. Each transistor is connected to a small number of other transistors, and each turns on or off depending on the state of the other transistors connected to it. In this analogy, the brain has neurons instead of transistors. Instead of the tiny wires that connect transistors together, neurons are connected to other neurons through junctions called synapses. Like transistors, neurons can turn other neurons on and off depending on the details of the connections between them. The idea, then, would be that some complicated wiring among the neurons in the brain performs computations much like a computer does, by building circuits.

Unfortunately, this analogy has some major limitations that miss what makes the brain really special. First of all, the brain is much bigger than a computer chip, with 86 billion neurons compared to only one or two billion transistors in today's computers. And each neuron is connected not to one or two other neurons, but to thousands—as many as 200,000.[2] This makes the system incredibly complex and hard to deconstruct. An individual neuron is also much more complicated than a simple transistor: while a transistor can only be on or off, a neuron can vary its firing speed to create additional complexity. Some neurons don't fire at all, but rather send out a continuously varying signal that affects how other neurons behave. Plus, there are many, many different types of neurons, each of which responds to incoming signals and sends out information differently from other types—even the same neuron's responses to signals can change over time. To cap it all off, some scientists estimate that your brain does all of this on about twenty or thirty watts of power,[3] which is less than most incandescent light bulbs.

It's also clear that the brain and a computer are good at different types of tasks. Computers are amazing at doing math, but it's not at all obvious how to make a computer display more "human" qualities, such as creativity. An iPhone can multiply billions of numbers every second, but asking, "Siri, tell me a story," will likely leave the user disappointed.[4]

The limitations of the brain-as-computer analogy should give some sense of the scope of the problem. Fortunately, through a combination of studying smaller portions of the overall brain, building up from the brains of simpler organisms, and developing new technologies that allow us to capture data about and to manipulate huge numbers of neurons, scientists are making progress.

Some of the success that scientists have had so far has come by figuratively breaking up the whole brain into parts and studying it in smaller functional units, both in humans and in other animals. We can try to get a foothold on the problem by looking at simpler systems to start to get a feel for some general principles about how brains work.

At large scales, the brain is highly organized—it is partitioned into regions that take primary responsibility for specialized tasks. For example, one can point to a region and say, "This part is the visual cortex, and that other part is mostly responsible for hearing." This organization of the brain into functional regions has allowed us to get a very broad idea of which regions do what, but that information only gets us so far. It's one thing to be able to point to a region and say, "This part handles vision," and another thing entirely to say *how* it handles vision. And while the visual system is actually relatively well understood, in many other cases we're still trying to figure out how the brain encodes information. Which neurons are involved in a person having a particular thought? Which are activated and which are turned off? And what do those patterns mean?

One example of an encoding that we are beginning to understand happens in a region of the brain called the entorhinal cortex. The then-husband-and-wife team of May-Britt and Edvard Moser shared part of the 2014 Nobel Prize in Physiology or Medicine in part for studying this region, and their work suggests that neurons in this part of the brain encode a fundamental concept: location.

During their studies, the Mosers discovered a peculiar pattern. If we were to record the activity from a cell in a certain part of the rat brain as the animal roams freely around its cage, we would see something strange. At first glance, it would look like this cell was firing at random. But if we were to take a picture of the rat's cage from above, then draw a dot wherever the rat was when that cell activated, a pattern would emerge. This neuron would turn on at defined places that form a triangular grid covering the floor of the cage; it would look a little like a beehive. From observations like these, the Mosers discovered a group of neurons that are now known as "grid cells," which define a grid over the rat's cage that presumably helps it keep track of its location. There are also other cells in the brain, known as place cells, that appear to fire when the animal is in specific locations. And it seems likely that these two types of cells are part of the encoding that allows a rat to know where it is.

It appears that similar encoding happens in humans. And some scientists speculate that since the Mosers' "grid cells" are some of the first cells to be lost to Alzheimer's disease, that may be why one of the first symptoms of Alzheimer's can be the person feeling lost.

We see similar correspondence between the outside world and the relevant parts of the brain in other contexts, too. For example, the human ear detects sound by splitting it up into frequencies, so there are some cells that detect high-pitched sounds and others that detect low-pitched sounds. These neurons are arranged in a row, with high frequencies detected at one end, middle frequencies in the middle, and low frequencies at the other end. In cases like this where we see a strong correspondence between the outside world and how the brain encodes information, we have made significant headway in determining what's going on in the brain.

But even with this early progress, it's sometimes hard to imagine that a three-pound tangle of cells can really do everything that the brain does. Is it really possible for the actions of a bunch of neurons to produce consciousness or intelligence? It's difficult to develop an intuition for how this might work, but one thing that might help is looking at one simple example of how even just a handful of components can do something kind of smart.

Imagine getting 9 people together to make a system that can learn. If each person just follows some insultingly simple rules, the group as a whole can display some sophisticated behavior.[5] And while this is not how the brain works, it might provide some inkling of how intelligent-looking behavior can happen once a few simple parts start working together.

The task for the group of 9 is to figure out whether a small image represents an X or not—a simple "yes" or "no" answer. Imagine this image is represented as a grid made of 9 squares. Each image always has 5 black squares and 4 white squares, but the arrangement is different every time. Here's the X that the network should recognize:

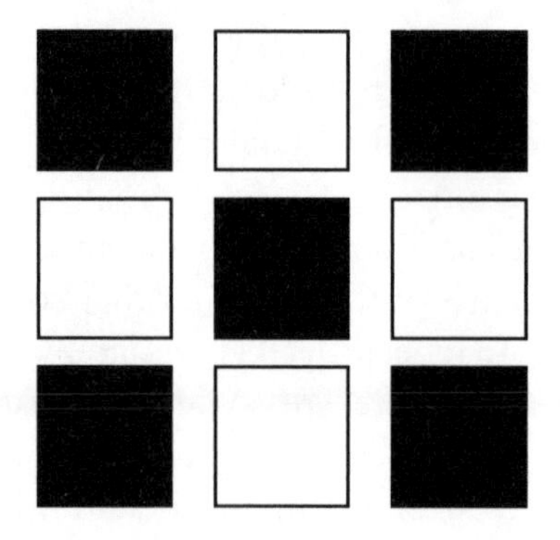

Figure 50: An X.

But while the network should be able to say that's an X, it should also be able to reject other patterns. For example, it should know that this pattern is not an X:

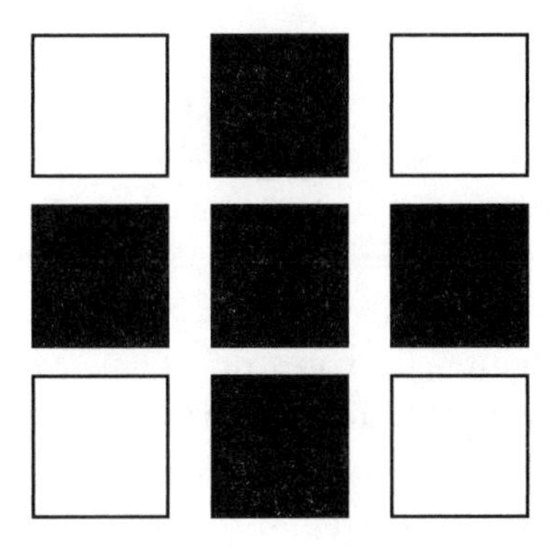

Figure 51: Not an X.

Even better, the group will achieve that recognition even though each person is allowed to see only one of the 9 squares. Consider the person who is assigned to watch the top-left square in that grid of 9. All that person sees is this:

Figure 52: A black square.

or this:

Figure 53: A white square.

Other than looking at their square, the only thing each person has to keep track of is one number—we'll call it their "influence"—that indicates how much their opinion contributes to the overall decision. (They can keep track of this number with coins or any other token to make sure no actual thinking is required!)

The group will then be presented with a grid. To make the choice, each person looks at his square. If it's black, he votes yes, and if it's white, he votes no. Then everyone makes a decision as a group by adding up the influence numbers of all of the people voting yes and comparing them to the influence numbers of everyone voting no. If the yeses have a larger number, the group guesses yes, and the group guesses no if the noes win.

Finally, the network is trained to recognize an X by adjusting the influence of each participant. We do this by using a few grids where we know the answer; we present a few X's and a few not-X's to train the network. For each training grid, everyone votes and we check the answer. If the group is right, nothing changes. If the group guesses it's not an X when it really was an X, every person adds 1 to their number if their spot was black. If the group guesses it is an X but it really isn't, every person subtracts 1 from their number if their spot was black. In either case, a person's influence doesn't change if they saw a white square.

And that's it. Now the group just continues going through all of the examples over and over again until the group gets the answer right for all of the examples—and, given certain assumptions, it will. The group will soon correctly be able to tell X's from not-X's, even though each person individually doesn't really know what's going on.

The secret is that the information about what an X looks like is stored in the "influence" number that each person is keeping track of. The training process gives people who are watching the squares that make up the X more influence, so when the group sees an X later on, those people dominate the vote. We can also think of those numbers (very, very loosely) as being like the strengths of connections between different neurons in a brain.

This exercise is actually inspired by the types of algorithms that computer scientists are using.[6] Researchers at companies like Google are taking inspiration from the brain's architecture to solve challenging problems that have eluded many other approaches. For example, researchers have used simulated systems of highly simplified "neurons" to read the zip codes people have handwritten on envelopes, to do speech recognition, to do stock market prediction, and to solve other problems that require sifting through huge amounts of data to look for patterns.[7] These types of systems have a surprising amount of power.

Of course, a simple task like recognizing numbers is a far cry from introspection or self-awareness, and as any researcher who has worked in machine learning can tell you, we still don't have a good understanding of how behavior like creativity might arise from a network of neurons. There have been a few bright spots, perhaps the most publicized of which was IBM's Watson, which demonstrated an ability to parse the questions on *Jeopardy!* while wiping the floor with the human champion, Ken Jennings. More recently, Google has developed a computer program that can describe what it sees in a picture. (It produces captions such as "A group of young people playing a game of Frisbee," or "A herd of elephants walking across a dry grass field," and it's right most of the time.[8]) Even this is light years away from a computer that has the imagination or creativity of a child. But given the surprisingly smart things that can be accomplished by a dozen people pretending to be neurons, imagine what 86 billion real neurons can do.

Even if we can build systems of artificial neurons that can exhibit some "smart" behaviors, it's an entirely different story to figure out how neurons actually work in a real brain. There's another wrinkle that makes all of this even more difficult: on the cellular level, there seem to be many different ways to build neural circuits that behave the same way. So how similar are two different, healthy brains? On a small scale, maybe not so similar.

A Cautionary Tale

It's not a huge surprise that some of the early advances in our understanding of the brain turned out to be more complicated than they looked initially. That wasn't much of a letdown, though, because no one ever really thought that we understood how any part of the brain worked on some basic level. Our early progress was mostly empirical—"And does *this* treatment help? It does? Great!"

For example, the Austrian physician Julius Wagner-Jauregg won the 1927 Nobel Prize in Physiology or Medicine "for his discovery of the therapeutic value of malaria inoculation in the treatment of dementia paralytica." Basically, he took people with dementia and infected them with malaria, and their dementia improved. This worked because those cases of dementia were caused by rampant syphilis attacking the brain; the malaria infection got the patient to run a fever, which improved the syphilis.[9] The "treatment" killed 15% of the patients who received it, but late-stage syphilis was almost always fatal. The best part? Penicillin was discovered one year later, in 1928. (Though it didn't become practical to mass-produce until the 1940s, when production was increased in preparation to treat the huge number of Allied casualties expected on D-Day.)

No one today is surprised that malariotherapy didn't stick around for very long as a treatment. Wagner-Jauregg probably never thought he had figured out some deep truth about how the brain works—he just thought he had validated a high-level hypothesis: that fever can help treat mental illness. Other parts of our understanding of the brain seem much more concrete, but even in those cases, there are some recent hints that even parts of the brain that seem very modular might be more complicated than we first imagined.

Take the foundational work of the neurophysiologists David Hubel and Torsten Wiesel, for example.[10] In the 1960s and 1970s, they used cats and monkeys to show how the visual cortex—the part of the brain that deals with the input from your eyes—processes visual inputs. They showed that cells that were right next to one another were processing light from adjacent parts of the visual field; that is, the center of one's vision and the part just to the right of that are interpreted by cells that are next to one another. Interestingly, they also found cells that detect edges—they are not turned on by light or dark, but rather by a boundary between light and dark. There were cells that would fire in response to lines of a specific orientation, cells deeper in the visual cortex that responded to right angles, and so forth. It looked like all of these cells that were processing vision were responding to features—like a vertical line, or a corner—in the visual field.

These results are still highly influential, but recent evidence suggests that the visual cortex may be a little more complicated than it seemed at first. For example, in rats that learned to expect a reward based on a visual stimulus, the cells in their visual cortex eventually grew to contain information about the expected reward[11]—that is, the cells were not just responding to edges or right angles or other features of what the rat was *seeing*, they also seemed to be responding to the *meaning* of what the rat was seeing. This is quite different from the simple model where the visual cortex simply processes the features of the visual field. The point is not that your visual cortex is involved in learning or anything like that, but rather that modularity, while a convenient assumption for a first pass at understanding the brain, isn't always as neat as we would like.

Eve Marder, a neuroscientist at Brandeis University, has thought a lot about how different two normal brains can be.

The first time I saw her speak, she was introduced as having "won every award that doesn't involve a trip to Stockholm." That's a fair overview, I suppose, but it doesn't really capture Marder's true influence on how we think about neural circuits.

Marder and her colleagues study a neural circuit in lobsters and crabs, as mentioned in Chapter 3. This circuit controls about 40 pairs of muscles that move teeth in the stomach; the rhythmic motions of these teeth help break down food. It's really nothing like a human stomach. The appeal of this neural system is that it strikes a nice balance between being interesting and being tractable. Marder says it is "complicated enough to have some of the interesting features of nervous systems, but small enough that you can record from all of the neurons that you need to at the same time." That made it ideal for study.

One of the questions that Marder's team has addressed is how widely two sets of neurons that behave the same way as a group might differ in their details; that is, they wondered if there is more than one way to produce the same behavior from a system of neurons. Indeed, in their experiments, two perfectly functional networks often differed dramatically in the details of how easily a cell fires or how strongly certain neurons interact with others. The behavior of the network was not precisely dependent on the characteristics of each neuron—instead, the overall behavior was robust to many changes, and there were some stiff combinations of neuron properties and some soft combinations.

The insensitivity to detail of these simple neural circuits suggests that human brains might actually be quite different from one another while still producing recognizably "brain-like" activity: personality, self-awareness, and creativity. Based on her work, Marder thinks the brains of, say, two healthy 20-year-olds with the same apparent IQ might still be rather different indeed. The details of how many cells are in one place, how strongly those cells connect with other cells, and which cells are connected, don't seem possible to specify exactly.

Marder sums it up this way: "All healthy brains probably follow the same rules, but all healthy brains are going to be individually different." And while that's an exciting notion, it also makes studying the brain that much more difficult: it reinforces the notion that the properties we care about come from many neurons working together rather than the details of any one specific neuron.

When studying the brain, scientists have to remember that humans' assessments of what's going on in their own heads can be incomplete or even inaccurate. Brains aren't always great at introspection. Of course, we have always known that our brains are doing a lot below the surface that our conscious minds aren't privy to. Imagine if you had to actively remember to breathe or make your heart beat, for example. But even high-level processes such as decision making may not be fully above the subconscious rumblings of the mind.

For example, scientists can show in the laboratory that the process of making some decisions can begin before the person reports being aware of having made a choice. In 2008, John-Dylan Haynes, a professor at the Bernstein Center for Computational Neuroscience in Berlin, Germany, and his colleagues reported the results of some experiments that involved measuring people's brain activity as they made choices.[12] They put volunteers into an fMRI machine—basically a big scanner that can image brain activity on a coarse level—and asked them to perform a task. Once they were in the machine, the volunteers saw two buttons, one on the left and one on the right, and a screen in front of them displaying various letters. They were asked to push either button at their leisure. After the person pressed a button, they were asked *when* they had made the decision about which button to press, and they answered by indicating which letter had been on the screen in front of them when they had made the decision. Usually this turned out to be about a second before the actual movement.

The researchers then went back and looked at the brain scans they had collected of these people while they were making their decisions, and they asked a simple question: based only on the brain scans, when could the researchers have figured out which button the subject was going to press? Amazingly, the scientists started seeing patterns of neural activity that could predict which button the subject was going to press as many as seven seconds before the volunteer even knew he had decided on a button. These patterns weren't perfect at predicting which button, but they were better than random; there was some information about the decision the person would eventually make in those brain scans.[13]

To be clear, these results do not mean that you have no control over your actions. Even these predictive patterns of brain activity can be controlled to some extent—for example, volunteers were able to succeed at a game that required them to "fool" a computer that was trained to detect the patterns of brain activity that tend to precede a button press.[14] But this study and many others like it underscore the point that our brains are doing a lot below the surface that our conscious minds are not always aware of. These findings demonstrate that the reality of cognition is even more amazing and complex than one might guess.

In addition to spying on the decision-making process, recordings from the brain can be used to allow people to interact with computers or robotic arms simply by thinking. For example, a 2015 study by researchers from Tsinghua University in China and the University of California, San Diego, used noninvasive electrical recordings of subjects' brains to allow them to spell words at relatively high speeds—up to about one letter per second—by simply looking at the letter on a screen.[15] The system detects the neural activity associated with this action and produces the correct letter.

In other cases, disabled patients have actually been able to control robotic limbs using their thoughts. For example, scientists were able to translate the brain activity of one quadriplegic patient into actions of a robotic arm.[16] The researchers first imaged the man's brain with an fMRI scanner and asked him to "imagine reaching and grasping." The brain activity they detected came from a region called the posterior parietal cortex, so the researchers implanted an electrical recording device into that part of the brain. The signals they recorded were able to distinguish among several imagined actions, such as moving a hand to the mouth, or rotating the shoulder. With this device, the patient could move the arm just by imagining the action he wanted to perform.

To allow a patient to spell a word or control a robotic arm, scientists had to associate certain patterns of brain activity with an action. In the future, this task of deciphering patterns of brain activity would be easier if we understood something about the language of the brain—how it encodes a particular idea, like "grabbing a coffee mug" or "the letter A," using the firing of its neurons.

Imagine that we were able to record the activity of all the cells in the brain at once—we can't even come close yet, but just imagine. What does the brain look like when it's thinking about the *Mona Lisa*, Immanuel Kant, the beautiful Platonic ideal of a triangle . . . or a celebrity? There have been a few tiny windows into this tremendously complex problem.

One such window came from a 2005 study by Rodrigo Quian Quiroga, director of the Centre for Systems Neuroscience and the head of bioengineering at the University of Leicester, and his colleagues at Caltech, UCLA, and MIT. These researchers recruited patients with epilepsy who have electrical probes implanted in their brains as a part of their epilepsy treatment. Those probes have a fortunate side effect: in addition to helping doctors to treat the epilepsy, the electrodes also allow researchers to record the activity of whatever neurons the probe happens to be nearby. Quian Quiroga and his colleagues used these probes to record the activity of a few neurons in each patient and test how those neurons responded when the patient was shown a series of pictures.

One neuron in one volunteer seemed to fire when the patient was shown a variety of pictures of Jennifer Aniston, but it didn't fire in response to pictures of other similar-looking actresses. A single neuron in another subject seemed to respond to pictures of Halle Berry, including pictures of her in costume as Catwoman. Even more impressively, it responded when the text "Halle Berry" was shown, but not when "Julia Roberts," for example, was shown.

One neuron alone probably does not "recognize" Jennifer Aniston, or Halle Berry. The encoding is likely to be more complex than that. Remember that the researchers could record from only the handful of neurons the previously implanted electrode happened to be near, so the odds of stumbling upon the one neuron that happens to encode the idea of Jennifer Aniston are astronomically bad. Instead, it's likely that each single neuron is a part of a more complicated encoding where a relatively sparse set of neurons fires to encode the concept of Jennifer Aniston. This single neuron might be part of a much more complicated pattern of neural activation that means "Jennifer Aniston."

Unfortunately, the techniques used in this study probably won't scale up to record from, say, all of the neurons in the brain. Additionally, it would be unethical to implant thousands of sensors into the brains of healthy people just to see how they think about celebrities. That's why new tools for studying the brain more comprehensively are necessary.

Professor Ed Boyden's work on the brain has the promise to deliver just the tools we'll need. Boyden is the head of the MIT Media Lab's Synthetic Neurobiology research group. To many people, it's not immediately obvious that a place called the "Media Lab" would be interested in doing "Synthetic Neurobiology," but having unusual interests is in keeping with the tradition of this place. Since its inception in 1985, the Media Lab has played host to a wide variety of work that didn't fall neatly into other categories. "The Media Lab has often been where MIT puts misfits who don't quite fit in with anybody else," says Boyden. Some of the first quantum computing at MIT was done here, back before it was an established field, and the e-ink display that you can find in an Amazon Kindle today got its start in the Media Lab. Even the Media Lab itself can seem out of place: it is still a part of MIT's School of Architecture and Planning.

Perhaps because of this culture, the Media Lab gave Boyden a chance when other places wouldn't. When he was first looking for a job as a professor, his idea for what he wanted to research—tools to help understand the brain—seemed a bit unconventional. "A lot of places didn't think neuroengineering was a viable career path," he recalls, "so I was turned down by, I guess, many universities I shouldn't name for an on-the-record publication." The problem was that making tools to study the brain was risky. "Nobody had ever made a career out of inventing more than one tool," he says. "Nobody had ever developed a whole series of things." Luckily, MIT gave him a chance, and Boyden's group quickly set out to develop that series of things.

Boyden is well equipped to develop new tools because, like many scientists who work with math and biology, Boyden has had training in many different fields, starting in high school. He skipped two grades, and participated in a program where he could take college classes in high school and get research experience. "I worked at a group that was trying to create life from scratch, which of course didn't work, or you would have heard about it," he says. He soon moved to MIT, where he initially worked on quantum computing. (That research, he says, "didn't work either.") After Boyden spent time at Bell Labs, he decided it was time for a change.

When deciding what to do next, he thought about one of his amateur interests. "I was very philosophically inclined," he says. He had spent a lot of his early college days "reading about literary criticism and semiotics and the representation of knowledge and, you know, existentialism," he recalls. "All sorts of stuff like that." These were appealing problems to him. "You sort of felt like you were wrestling with reality and the nature of existence and so on." He liked philosophy, but didn't want to do it for a career. It did lead him to where he is now, though: at that time, "it felt to me that the next phase was really the brain," he says.

And so Boyden became a neuroscientist. Luckily, he had a fellowship from the Hertz Foundation, which affords enormous freedom. They're generally pretty happy to let fellows follow their instincts. "At the time, the Hertz Foundation was not known for funding biology," he says, so telling the foundation that he wanted to do biology could have been a bit awkward. "I called them up and said, 'I'm gonna quit doing quantum physics, can I take my fellowship to go to neuroscience?'" he remembers. "And they said, 'write a letter to the board explaining your decision.'" A few weeks later, he was approved.

One of Boyden's first contributions to neuroscience was his work on a technique called optogenetics. The general idea was to turn any arbitrary set of neurons in a brain on or off simply by shining a light on them. But that's easier said than done. "The hard part is making the neurons controllable by light," Boyden says, "because neurons don't normally respond to light," except in some special cases. He and his collaborators decided to try to make this happen by finding a protein that can turn visible light into electrical current—which is what drives neural activity—and inserting it into neurons. "All sorts of organisms have these light-sensitive molecules," says Boyden, so it was a matter of finding the right protein. After searching a number of scientific papers, they found one from algae that they thought might work. Light-sensitive protein in hand, it was time to see whether they could actually control neurons with this strategy.

With his collaborator Karl Deisseroth, a professor of bioengineering and of psychiatry and behavioral sciences at Stanford University, Boyden gave it a shot. "Karl transfected the genes into the cells, and I did the recordings and delivered blue light," he says. And at one A.M. on August 4, 2004, Boyden turned on the light and saw a response from the cells. When he shined the light on the neurons, they "turned on" by producing electrical spikes. Optogenetics was working.

The final step before using this in a live animal was the challenge of installing this light-sensitive protein in the neurons. Luckily, the researchers could hijack a virus to do the work for them. "There's a virus called AAV that basically all of us have—it doesn't cause any symptoms," Boyden says. AAV normally inserts itself into the genome of the cells that it infects, but we can reprogram it to instead insert a gene of our choice into the cell. In this case, they built a version of AAV that inserts the gene that tells the cell to make the light-sensitive protein. They could even make it so that this light-sensitive protein turned on only in the cells they wanted it to be active in. This technique now allows us to turn on or turn off specific neurons in the brain with just a flash of light.

Seeing this happen in a live animal with a fully functioning brain is really amazing. You see a mouse with a fiber optic cable sticking into its skull just wandering around, acting normally. All of a sudden, the cable glows blue and the mouse immediately starts running in a circle. The light turns off, and the mouse stops as if nothing ever happened.

With optogenetics, scientists can now poke at individual neurons and see how they affect the behavior of the overall system, the brain. For example, researchers can systematically stimulate different regions of the fruit fly brain to see which regions cause reproducible behaviors when turned on,[17] or even "steer" small worms by specifically activating some of their neurons, causing them to turn left or right.

Optogenetics is a fantastic tool, and Boyden's group has been working to make this tool freely accessible to any lab that wants to use it. "There's just too much to do for one group," he says, so the collaborative spirit runs high here. And Boyden knows that there's a lot more yet to do on the way to understanding the brain. "Controlling a cell is only one part of the puzzle," he says. "Ideally you can read all the neurons in the brain and you can also map all the connections and the molecules in the brain." That's an ambitious goal, to be sure, but they and many other groups are working on it. President Barack Obama's BRAIN (Brain Research through Advancing Innovative Neurotechnologies) initiative provides a lot of funding to develop more tools that will hopefully one day help us decipher what's happening. "It's about building tools that help us understand the brain," Boyden says, "and I think that's a really important emphasis because the existing tools *don't* solve the brain." He sits forward in his chair. "And without new tools, how do you make the right measurements?"

One other new tool Boyden's group is using is a microscope they and their collaborator Alipasha Vaziri built that can take 3D images of brains. It uses more or less "the same tricks that our eyes do," he says, and utilizes techniques that are common in some other parts of science and engineering. "We have two eyes and each eye takes a slightly different picture of the world, and so we can see in three dimensions," Boyden says. This 3D microscope just uses extra lenses to work a little like extra eyes, and the team can then reconstruct the 3D image with a computer. "We've now basically made a microscope that has multiple eyes," he says. "We went on to image all the neural activity in the brain of a small vertebrate, the larval zebrafish. This animal has 100,000 neurons, and we gave the animal an odorant and then we isolated the 5,500 neurons computationally that were activated by this smell. So this idea that you could image an entire neural circuit is now possible."

But we have a long way to go yet. "I think we don't know much about the brain," Boyden says. Consequently, one of his current focuses is devising ways to study the brain that don't assume anything about how it works. After all, we're constantly discovering new things about the brain, and we'd like the data we collect today to be useful tomorrow, even if one of those new things turns out to be important. His lab recently published a paper describing a new technique called expansion microscopy that might be useful in that respect;

expansion microscopy physically expands tissue samples so scientists can take higher resolution pictures of them.[18] The sample swells up much like a wet diaper does. Boyden thinks this procedure will allow his lab and others to map the locations of the different molecules in the brain at high resolution.

For Boyden, part of the appeal of avoiding assumptions when studying the brain is grounded in historical precedent. "In neuroscience, there are many examples of people who thought they were doing well, and maybe even the whole world thought they were doing well, but they weren't," Boyden says. "In fact, if you look at the first century of Nobel Prizes in neuroscience, a significant fraction of them I wouldn't say were awarded for *mistakes*, but, you know, people look back and wonder, well what was going on there?" He hopes to avoid this fate by trying to make sure that what we think we know doesn't color the data collection process. "We want to stop making assumptions," he says. "We want to not wake up one day and find out we've been wasting our time." Based on his progress so far, I don't think that's something Boyden has to worry about.

Simulating the Brain

The Human Brain Project is a massive effort by scientists in the European Union to study the brain. The project will cost more than a billion euros over ten years, and it involves researchers from more than 100 institutions in 2 dozen countries. There are 13 subprojects within the overall initiative that include collecting new data about mouse and human brains, building computer software that combines various kinds of data, and developing computer simulations of the brain.

Sean Hill, a leader of one of the Human Brain Project's subgroups, has been working with neuroscience for a long time. He says he's always been interested in both computers and biology: "I did an undergraduate degree in the U.S. at kind of an experimental undergraduate place where you could make up what you do," he says. For him, that meant recording and building mathematical models of neural circuits. He's also an IBM alum, so he has the computer chops to match his interest in biology.

Hill is the codirector of neuroinformatics for the Human Brain Project, and he is responsible for one of those 13 subprojects within the overall initiative. His team is focusing on building the infrastructure that will allow researchers to combine data from many different techniques and places into one representation. "This is the platform that is responsible for making it possible to integrate data from lots of different data sources into brain atlases," Hill says. Just like a normal atlas incorporates all different kinds of data—national borders, roads, rivers, buildings—into one common map, brain atlases are an attempt to represent these data in a common way. "That's really what our core mission is, to provide that framework and then to show how it can work," says Hill. The goal here is to make data about the brain accessible and usable by the research community. For researchers who work on the brain, "this is their search engine," says Hill, "to find the data and find it in a way that they can in principle, make use of it." Fundamentally, Hill says, his team is trying to provide a common framework for brain researchers to use and build upon.

These unified data sets will then inform computer simulations of the brain; another subgroup within the larger project is using these integrated data sets to run simulations of parts of brains. This is a wildly ambitious goal, but Hill seems to have a very realistic attitude. "The only possible way you know that the model has any value at all is the extent to which it replicates multiple experimental findings," Hill says. "We're not manually tweaking parameters to elicit a particular network behavior," he says. Using the data they have, they model neural circuits as best they can, Hill says, "and then we see what emerges."

In some instances, what emerges is behavior that at least looks a lot like what you see in real systems of neurons. Some of the behaviors they look at are hard to understand without a background in neuroscience, but others are fairly simple. For example, if you take a slice of one part of the brain under certain conditions, Hill says, "you can get low-frequency oscillations," at a rate of about one per second. "Now that's very, very baseline, very basic behavior," he notes, but this is one example of a behavior that any realistic brain simulation should account for. "This is well known and well studied in many different preparations." Then, under another set of conditions, the behavior changes to something more sporadic. "It responds to stimuli totally differently," he says, in a way that's a bit more like what happens in a live animal. The project's simulations can account for these kinds of observations.

Despite the project's ambitious goal of eventually modeling a human brain, Hill is keen to point out that these models will just be first drafts. "We're probably never going to have a perfect model of any brain," he cautions. Instead, he thinks we should think of this as "a tool for organizing our data and our knowledge to make predictions that can then be tested and can be supplemented by additional data." In other words, he and his team hope to make models that can be continuously improved as we understand more. "What we're doing is we're building a facility, the capability, and a first draft brain model." This, he admits, "is still crazily ambitious, but it's something that we can see is possible, given what we know."

In the meantime, Hill's group is focusing on building the tools to let brain scientists see what their data and ideas might mean. Ultimately, Hill sees this group of tools as a way of helping us improve our understanding. "We know there's going to be lots that we don't know," says Hill, but "we think modeling is the key tool to understanding." By incorporating many different kinds of data and continuously refining their ideas about how the brain works, Hill and his colleagues hope to use these tools to make sense of what's going on. "It's a tool that I think is critical to developing a real theory of the brain," says Hill. In the end, they're building models not because they already understand it. "We're building it in order to understand."

CHAPTER ELEVEN

Death and Taxes

Aging Is Governed by an Organism-Wide System that We Might Be Able to Manipulate

Nearly all organisms break down as they get older. Humans, for example, become increasingly frail: our strength diminishes, our skin wrinkles, and our wits become slower. Similar signs of aging are also manifest in the millimeter-long worm *Caenorhabditis elegans*, though over a much briefer period. During its two to three weeks of life, the worm gets discolored and wrinkled. Eventually, its crawling becomes uncoordinated, it stops feeding, and it dies. This short lifespan makes it practical to study aging in the laboratory using the worm as a model.

We often think of aging as being inevitable—it's said that the only two sure things are death and taxes—but there's no fundamental physical law that says organisms must break down as they get older. Indeed, there are many interventions that can considerably extend the lifespan of organisms scientists study in the lab. And piecing together how those organisms are able to modulate their own aging may help us better understand the process in humans.

Many of the known methods for extending lifespan are intimately connected with the animal's metabolism and its perception of how much food and energy is available. For example, scientists have known since the 1930s that a reduced-calorie diet can extend the lifespan of many living things.[1] In rats, fruit flies, and yeast, restricting the organism's intake of food makes it live up to 40% longer, depending on the organism and the experiment. In worms, this effect was first described in 1977.[2] Calorie restriction doesn't seem to work in all organisms, though. It is not yet known for sure whether calorie restriction extends lifespan in humans, but the evidence in monkeys is mixed,[3] and the limited types of studies we can do on people suggest that being of average weight is best.[4] Exercising and eating healthy food in moderation is probably the best bet.

Another way to alter the lifespan of *C. elegans* emerged in the late 1970s when Michael Klass, a postdoctoral fellow in David Hirsh's biology laboratory at the University of Colorado in Boulder showed that temperature affects the worm's lifespan quite drastically.[5] At high temperatures of 25°C (77°F), worms live for a little over a week, but they live for three weeks at 15°C (59°F). These changes in temperature would be expected to alter many of the worm's internal processes because worms are "cold-blooded"—scientists prefer "poikilothermic"—which means that they do not maintain a constant internal temperature.[6] A change in temperature alters the speed of the chemical reactions that are necessary for life, and a shift from 15°C to 25°C affects each reaction differently. Therefore, worms must actively respond to temperature changes, including by making changes that impact their lifespan.

Indeed, worms adapt their lifespan to match the temperature at which they are living; a normal worm that grows to adulthood at a high temperature and is then moved to a low temperature lives much longer than a worm that spends its whole life at the high temperature. It seems that they have proteins that can sense colder temperatures and send a signal that can adjust the animal's lifespan accordingly.[7] They also "remember" the temperature at which they grew up and, if given the option, tend to move toward regions held at that temperature.

In the late 1980s, several scientists started to discover that lifespan can be extended through changes in the worm's genes, too. One group at the University of California, Irvine, the biologists David Friedman and Thomas Johnson, looked at thousands of mutated worms and found one mutation that increased the lifespan of worms by 65%.[8] This result was exciting, but their worms had another feature that suggested a relatively trivial explanation for the increased longevity: reduced fertility. Since producing offspring requires significant energy, scientists at the time hypothesized that this reproductive energy might come at the expense of maintenance of the individual. Consequently, it was thought that lifespan extension accompanied by fertility defects was only of moderate interest. Though we later learned that the fertility defects were a side effect of another, separate genetic alteration in those worms, the loss of fertility somewhat tempered scientists' excitement at the time.

Just a few years later, Cynthia Kenyon, a biologist at the University of California, San Francisco, and her colleagues isolated a strain of worms that had a mutation in the gene *daf-2* that caused them to live twice as long as normal worms.[9] Even better, this team showed that reproductive defects probably weren't the source of the longevity by destroying normal worms' egg-producing cells with a laser and showing that they didn't live any longer than normal. This mutation, it seemed, was doing something exciting.

Today we know that *daf-2* and other related genes are a part of the insulin-signaling pathway in worms. Insulin is a small protein that acts as a hormone: it travels around the body and carries a signal to many different tissues and cells. In humans, insulin tells cells to take up glucose from the blood, as discussed in Chapter 4. In worms, insulin plays an important role in the regulation of energy usage, growth, metabolism, development, and lifespan. The genes Friedman, Johnson, and Kenyon identified are involved in sensing insulin and turning genes on and off to respond to that signal. Reducing insulin signaling in adult worms seems to lead to a longer lifespan by turning on genes that help the worm prevent or repair damage, by turning on the worm's defenses against bacterial infection, and by turning off genes that tend to shorten the worm's lifespan.

Insulin helps to link all of the parts of the worm together and to transmit signals that affect lifespan. For example, insulin signaling involves communication from neurons to other parts of the body. Turning on insulin signaling in a worm's neurons but not in the rest of its body is sufficient to give it a normal lifespan.[10] These neurons then communicate with the rest of the cells in the worm in order to regulate how those other tissues age.

One other way the worm's neurons can communicate with other tissues to impact aging is by using a small molecule called octopamine,[11] which is closely related to adrenaline. Octopamine signaling appears to affect the integrity of muscle cells' mitochondria, which are often called the "power plant" of the cell because they produce most of the energy that the organism uses. They contain specialized machinery that can efficiently extract energy from the cell's food. In the worm's muscles, the mitochondria normally exist in clean stacks that look like a pile of pancakes, but as the worms age, those mitochondria become fragmented and misshapen:

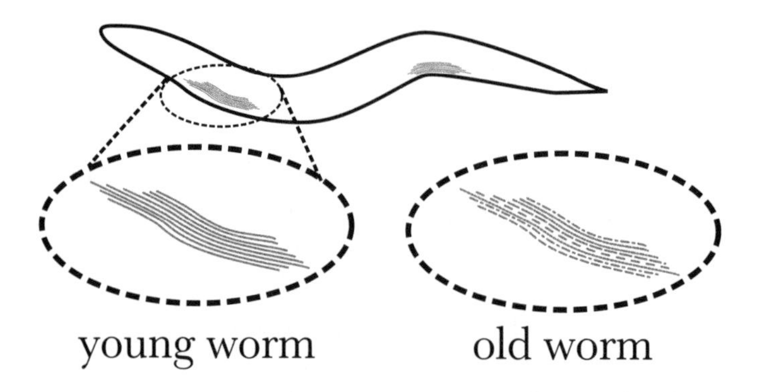

Figure 54: A worm's muscle mitochondria break down with age.

This mitochondrial breakdown is governed in part by signals from neurons that sense the overall amount of energy available. When energy stores are low, octopamine signaling is turned off and the mitochondria tend to stay healthy. But increased octopamine signaling in response to plentiful energy tends to cause mitochondrial fragmentation.

This regulation of aging happens across whole worms, but it occurs at a smaller scale, too—inside single cells. In yeast, a single-celled organism that displays some symptoms of aging similar to those seen in more complicated organisms, mitochondria break down just as they do in worms. In 2012, scientists at the Fred Hutchinson Cancer Research Center in Seattle published the results of their work on this process. Postdoctoral fellow Adam Hughes and principal investigator Daniel Gottschling found that the cell's mitochondrial problems could be delayed by protecting the integrity of another part of the cell, an acidic compartment called the vacuole.[12] The yeast vacuole performs many functions, including storing the building blocks of proteins, breaking down existing proteins, and helping the cell maintain the proper internal environment. When the vacuole starts to lose its acidity, though, the researchers observed that the mitochondria soon start to break down. By artificially increasing the acidity of the vacuole by forcing the yeast to make a protein that pumps protons into it, the researchers made the yeast's mitochondria last much longer. The mitochondrial breakdown might happen in part because the vacuole is no longer properly storing some protein building blocks, though this connection is still an area of active research.

Mitochondria communicate with the rest of the cell in mice and humans, too. For example, mitochondria participate in the signaling involved in cell death and in recycling old cellular parts for reuse.[13] It was once thought that mitochondria were more or less interchangeable sources of energy for the cell—you wouldn't expect much of a difference if you swapped a Duracell for an Energizer in your calculator. But swapping out one type of mitochondria for another type appears to have effects on learning and cognition in mice,[14] which suggests that it may be possible to have mitochondria that are "mismatched" with the rest of the cell. And in humans, changes in the mitochondria are risk factors for conditions such as Alzheimer's disease and schizophrenia.[15]

While the actual process of aging plays out during adulthood, some of the events that can affect a worm's lifespan happen long before that worm is even born. Eric Greer, a principal investigator at Children's Hospital Boston and assistant professor at Harvard Medical School, found one such effect. While a graduate student at Stanford in the lab of professor of genetics Anne Brunet, Greer showed that a genetically normal worm's lifespan can depend on something that happened to its great-grandparent. That is as if your daily jog could give your future great-grandkids a few extra years of life.

Greer started out by looking at proteins that add and remove marks from histones, the proteins that DNA is wrapped around. He systematically got rid of each of these proteins and observed the effects on the worms' lifespans. He found that when the worms were missing one of these genes, they lived 20 to 30% longer.[16]

Greer then designed an experiment to see whether this effect might be heritable; "It was kind of a shot in the dark," he says. In this experiment, an initial generation of worms would have one of these mutations that led to increased lifespan, but their kids would not have the mutation.[17] One might expect the lifespan of these genetically normal offspring to return to its usual length, but that's not what happened. Even though the initial worms' descendants had normal DNA, they still lived longer than usual: the first-, second-, and third-generation kids of those long-lived parents also lived longer. The increased lifespan was inherited for three generations, even though the gene that Greer initially got rid of in order to make them live longer was back to normal. How this works is still under study.

Humans have a major advantage over worms when it comes to combating aging: worms cannot replace damaged cells, but we can. All of a worm's cells stop dividing once it reaches adulthood, so it is stuck with the same cells for the rest of its life. Humans, in contrast, have special cells—stem cells—that can divide to replace old or damaged cells, but not indefinitely.

When scientists first started to grow cells isolated from whole organisms in a petri dish, they discovered that these cells could grow for only a limited time. The usual procedure for culturing cells is to put them in the dish with some nutrients and let them grow. As long as they have the proper environment, certain types of cells are happy to grow in a dish. But in the early 1960s, Leonard Hayflick, then a researcher at the Wistar Institute in Philadelphia, noticed that there seemed to be a limit to how long this could continue for any given group of cells.[18] After about 40 to 60 cell divisions, he noticed that the cells stopped growing and entered a non-dividing state that would come to be called senescence. This point at which the cells stop growing is known as the Hayflick limit.

It turns out that many cells, including Hayflick's, keep track of roughly how many divisions they have undergone. Each of the chromosomes in a cell has a "cap" on each end called a telomere; the telomere protects the rest of the DNA, like the plastic tips on shoelaces. When a cell divides, it loses a small part of its telomeres, and when the cap wears down enough, the cell stops dividing and enters senescence.

Scientists believe that cell senescence is one important part of human aging. The cells that normally divide to replace old or damaged tissue in a young and healthy person eventually lose their ability to proliferate. This increased senescence with age can limit a person's ability to replace damaged tissue.

On the other hand, senescence also probably serves an important function: keeping cells in check. Uncontrolled cell division is one of the hallmarks of cancer, so "cutting the brakes" on senescence entirely probably isn't a good idea, either. Still, some evidence suggests that "pruning" senescent cells by selectively killing them can reduce some problems associated with aging in mice and can lengthen their lifespan,[19] so some scientists hope that we might one day be able to reduce the effects of aging in humans with a similar strategy.

Perhaps when looking for ways to improve our ability to replace damaged tissue, we should take our cues from animals that are much better at regenerating and healing than we are. Some amphibians, including newts and axolotls, can regrow entire limbs. Tiny flat organisms called planaria that are commonly found in lakes and rivers are sometimes called "immortal under the edge of the knife" because they can regenerate lost body parts seemingly without limit. Some research even suggests that planaria can somehow retain simple associations of food with light after they are chopped in half and regrow.[20]

These organisms are different enough from humans that it is unlikely we would be able to harness similar regenerative abilities directly. But studying them might teach us about how these animals are able to mobilize their cells to replace huge portions of the adult body.

Even though we still have much to learn about how aging works in humans, there are already dozens of companies that are trying to use what we know to develop drugs and other therapies to help people. There are plenty of antiaging scams on the market, but there are also a few big names who are trying quite seriously to tackle aging. Harvard professor of genetics David Sinclair, for example, cofounded Sirtris Pharmaceuticals, which hoped to design pharmaceutical drugs to combat aging. Despite some controversy over the validity of the science on which the company was based, GlaxoSmithKline (GSK) purchased Sirtris for $720 million in 2008. GSK shut down Sirtris's original location in 2013 but said at the time that the Sirtris compounds would continue to be developed. Also in 2013, Google helped found a new company called Calico—the California Life Company—to study aging, and they soon made waves in the field by hiring several prominent researchers. And a company called

Alkahest, founded in part by professor of neurology Tony Wyss-Coray of Stanford, is trying an unusual strategy for combating Alzheimer's disease: transfusing elderly patients with blood from young donors.

The basis for the young blood trial began with laboratory experiments using a technique called parabiosis. In parabiosis, two mice are joined together surgically so that their bloodstreams become linked. Since the joined mice share the same blood, anything that's floating around in their blood will also be shared—so any signals that are carried in the blood will affect both individuals.

When researchers joined an older mouse with a younger mouse, some parts of the older mouse that had previously withered with old age seemed to have improved.[21] Scientists have found that these "rejuvenated" mice have healthier-looking hearts, better liver function, younger-looking brain cells, and even a better sense of smell than old mice that were joined to other old mice. It seems that something in the young blood has beneficial effects or that the younger mouse scrubs something harmful from the older mouse's blood. Researchers are still investigating how this happens.

Even in the absence of a detailed understanding of the rejuvenation effect, Dr. Wyss-Coray's company decided to go ahead with preliminary tests in humans. Alkahest is performing a small clinical trial in which patients with Alzheimer's disease will receive transfusions of young blood and be monitored for any cognitive improvements. This is possible due to the nature of the intervention; we've been doing blood transfusions safely for decades, so it's easier to justify trying it out in humans. The study is small and limited in scope—and is probably a long shot—but a positive result would attract a huge amount of attention and money. The results are expected sometime in late 2016.

CHAPTER TWELVE

Your Microbiome and You

The Body Is Host to Trillions of Microbes that Affect Human Health

Collecting fecal samples presumably wasn't anyone's idea of a fun time, but a team of scientists from Washington University in St. Louis and the University of Malawi had a job to do. The researchers were taking samples from Malawian children because they hoped that the data they could glean from testing the samples held the key to understanding kwashiorkor, a devastating type of malnutrition. Malnutrition is a huge problem in Malawi—about half of Malawian children under five experience stunted growth from inadequate nutrition[1]—but the magnitude of the problem likely wasn't the only reason the team was working on this issue. It also seemed that kwashiorkor was a type of malnutrition that was affected by more than just what the child ate.

The scientists were studying hundreds of pairs of Malawian twins, and these children displayed an odd pattern. In almost half of the pairs of twins, one twin was malnourished and one was healthy. This seemed strange; after all, identical twins have almost exactly the same genes, and fraternal twins are genetically very similar. When they also are raised by the same parents, live in the same house, and eat the same food, many of the typical environmental and genetic variables that substantially impact a child's health are very similar for both twins. When all of these risk factors are the same, you would probably expect that both twins would be healthy, or both would be unhealthy. So what was going on?

Maybe kwashiorkor isn't only about how much food you get, but also about how well you use that food. If both twins are eating the same things, but the healthy twin is simply able to extract more nutrients from that food than her sister is, then that would explain the discordance. The scientists suspected that the efficiency with which these children were able to extract nutrition from their food might be influenced by the types of bacteria living in each child's gut. If a child has bacteria that allow her to efficiently extract nutrients from her food, the thinking goes, she can stay healthy on a diet that would lead to malnutrition in someone with "inefficient" bacteria.[2]

To study this problem, the researchers first gave all of the children an energy-rich paste known as ready-to-use therapeutic food, or RUTF, and monitored how they responded to the treatment.[3] At first, they did well: they gained weight, became healthier, and the bacteria in their guts perked up immediately.[4] The microbiomes of the malnourished twins started to exhibit signs of metabolic activity that are generally associated with healthy children and "efficient" bacteria. Everything was going well . . . until the RUTF was stopped. When the researchers finished treating the twins with supplemental food, each child returned to his previous state of either malnutrition or general health, and his bacteria also settled back into their old patterns. The children couldn't be cured for the long term with RUTF; they improved only while they were getting this energy-rich food.

But what about those bacteria? They seemed to be associated with the children's health, but that doesn't necessarily mean the bacteria were involved in *causing* the malnutrition—they might just be a symptom. So the scientists wanted to know: Were those bacteria actually part of the problem? To find out, the team needed some poop.

Before starting the twins in their study on RUTF, the scientists had taken stool samples from each of the children to understand what bacteria were living in them. About 60% of the dry mass of your feces is actually bacteria from your gut, so this was an effective way of getting at the microbes that the children were carrying. The scientists then took these stool samples and used them to inoculate mice that didn't have any microbes in their own guts; that way, the scientists knew that whatever microbes grew inside these mice came from the Malawian children.

Once the transplanted bacteria had taken hold in the mice, the researchers fed them food that was intended to mimic the typical Malawian diet, and they watched to see if the mice would develop malnutrition. The animals who got bacteria from the malnourished twins became malnourished, and the mice who got bacteria from the healthy twins remained healthy—even though both groups were eating the same amount of food! And recent work has shown that introducing certain species of bacteria into the guts of the malnourished mice can prevent the growth impairment they would otherwise experience.[5] At least in mice, it seems that the particular types of bacteria that live in the gut can actually be one of the factors that causes these developmental consequences. And as scientists are learning, it turns out that these tiny microbes affect a lot more than just malnutrition.

Most people are already used to thinking of parts of the human body as systems, such as the nervous system, the cardiovascular system, or the immune system. However, there's one important system that few people know about, a system that accounts for about half of the cells that actually make up the body. It's the part of the human body that isn't human.

A typical adult person is composed of about 40 trillion cells, but he or she is host to at least that many microbes living on and in the body.[6] These microbes are mostly bacteria, but there are also plenty of less-familiar single-celled organisms. All told, these microbes make up about three or four pounds of a typical person's weight. Most of them live in the colon and digestive tract, though they also inhabit the skin, ear canal, lungs, and any other space where they can find a hospitable environment. These creatures are constantly growing and living and fighting and dying within us, and most of us have no idea.

All of the bacteria and other microbes that live in and on you are collectively called your microbiome,[7] and they have historically been largely ignored as a factor in human health. An explosion of recent research, however, has shown that these microscopic organisms—like the microbes that contributed to the twins' malnutrition—actually play a huge role in your life.

Most of the time, the bacteria of the microbiome are our friends: they're our partners in digestion and nutrition, and they can help fight off foreign microbes. "Most people have this view of our encounters with microbes from the perspective of disease. But that couldn't be farther from the truth," Dr. Jeffrey Gordon, the scientist who led the study on malnutrition, told National Public Radio.[8] Most of the bacteria that live in us are symbiotic; that is, they play nicely with our bodies and provide some benefit in exchange for food and a place to live. In fact, the body needs them to be there, and it sometimes even promotes their growth. A mother's breast milk, for example, contains complicated sugars that the child alone can't digest, so they may be there to favor certain microbes in a developing child. In short, think of the microbiome not as a group of invaders, but rather as friendly neighbors—albeit neighbors who live in your basement and eat your table scraps.

If the normal patterns of the microbiome are disrupted, though, things can get ugly. We're still not sure exactly what a "good" or a "bad" microbiome looks like—if it's even possible to categorize them that way—but it's clear that some bacterial populations can have negative side effects for their human host. If your microbiome contains certain types of microbes in certain proportions, they may affect your weight, your likelihood of getting cancer, and maybe even your mood.

We acquire our microbiomes quite early in life; the general composition of the gut microbiome is established stably before we're about two years old. The microbes come from many places, including the environment we are exposed to as young children, the food we eat, and even our mother's vaginal canal. In fact, babies who are delivered by cesarean section have significantly different microbiome compositions from those who are delivered vaginally, though it seems that these differences usually disappear with time. Once the composition of the microbiome is settled, it seems to be relatively stable, though events like changes in diet or courses of antibiotics can certainly change those patterns.

The microbiome mostly eluded notice until recently, in part because it's difficult to study. The microbiome is a complicated system, so it's hard to know how to dissect it piece by piece to research it. Each mini-ecosystem within your body contains a distinct microbial profile, and those profiles can change abruptly with time and life events. Even worse, many of the species that live happily in the body simply won't grow under the typical conditions scientists use to culture bacteria in the lab, so we're missing most of our normal tools. It's like trying to figure out how a clock works by taking pictures of it from a distance and hoping that it doesn't spontaneously change into a toaster while you're trying to study it.

Luckily, techniques from systems biology provide a way to attack the problem. We don't have to understand everything about every species of bacteria in the gut in order to study the properties of the microbiome system. In fact, we can make a lot of progress by simply knowing which bacteria are present in what proportions and how those bacteria interact with each other. By studying how the system as a whole reacts to perturbations, we can learn a lot about how the microbiome impacts the rest of the body.

When I first began thinking about the microbiome, I decided to start with my own. Finding out what's living in one's gut requires special laboratory equipment that most people don't have access to, but there are companies that will analyze anyone's microbiome for a pretty reasonable price.

Most of them work by sending you a sample collection kit, which you then use and mail back for analysis.

The company I used is μBiome (pronounced "You Biome"), run by an energetic entrepreneur, Dr. Jessica Richman. In 2012, she and two graduate students at the University of California, San Francisco, who were studying the microbiome became frustrated that only a select few people could learn about their own microbial roommates. "The only people who could do it were maybe 100 researchers around the world who were microbiologists," she says. To work with the microbiome, you had to be a professional scientist, but Richman and her cofounders began to wonder if there was a way to change that. Their answer was to start μBiome.

To get the money they needed to launch the company, Richman and her cofounders turned to a distributed financial model called crowdfunding. They submitted a business plan to an online funding platform called Indiegogo—you may have heard of a competing site, Kickstarter—where thousands of people around the world can pitch in small amounts of money to make a project happen. They didn't know how well it would be received, but the timing was right, and the campaign took off. Richman chalks up this success to a "groundswell of interest" in the microbiome around the time of the campaign, but the results were phenomenal. The campaign ultimately raised $350,000, far more than its $100,000 goal, and got about 3,500 people involved in the project.

Using μBiome's kit, I sent them a swab of bacteria from my gut. They then used simple laboratory techniques to look for a very specific genetic sequence in the sample that is common to nearly all bacteria, but has slight variations in each different type. By determining how much of each different variant of this sequence was in my sample, μBiome deduced which types of bacteria were in my gut and in what proportions. I got an e-mail when my results were ready, and I took a look online. According to μBiome's data, I have more bacteria from a very broad group called *Firmicutes* and fewer bacteria from the group *Bacteroidetes* than the average person who sent in samples. But those numbers don't really mean much on their own, since those bacterial groups are too inclusive. (For comparison, the analogous group for humans—vertebrates[9]—also contains blue whales and peregrine falcons.)

When I drill down to the results for more specific groups, I find that some of them seem to differ significantly from most of the people μBiome has tested. For example, I seem to have a ton of a group called *Blautia* (13.4% in my sample compared to the average of 7.55%) and *Bifidobacterium* (6.11% in my sample compared to the average of 0.883%). These narrower groups are a little more informative, but grouping at this level is still the equivalent of lumping together a dachshund and a dingo—with no way to determine which was the cute hot-dog kind of canine and which was the kind that might eat a toddler.

Tests like μBiome's can also provide some limited guesses about what kinds of general functions those microbes might perform.[10] For example, μBiome predicts that my gut bacteria as a group might be good at breaking down the building blocks of proteins but relatively bad at metabolizing a type of omega-3 fatty acid. A lot of current science about the microbiome is centered around understanding what kinds of jobs a particular set of microbes can do and how they interact, and μBiome isn't set up to answer those kinds of questions in much detail. The results of the test are more of a bird's-eye view of one's microbiome than an exhaustive exploration.

For now, if you want to get your own microbiome analyzed, do it for fun—not to make any life decisions. But if the spirit behind μBiome has its way, we will eventually routinely get information about the bacterial part of our bodies that we can use to improve our lives and our health.

Unlike in Malawi, malnutrition isn't typically a primary concern for most people in developed countries. Obesity, on the other hand, *is* a problem. About one quarter of adults in the United Kingdom and more than a third of adult Americans are obese, and weight problems are a huge risk factor for everything from heart disease to type 2 diabetes. For very obese people who haven't been able to lose weight on their own, one potential intervention is bariatric surgery—surgery to aid weight loss. And for dangerously overweight people, this surgery can be a lifesaver.

Sam, who ran a restaurant in Philadelphia, was one such person. "I'm in the food business; I've been trying to lose weight for thirty years," he tells me, but he hadn't found success. Then the situation became urgent. "I needed a kidney transplant, but my BMI was too high," he explained. The choice was simple: lose weight, or go without a kidney. A doctor recommended bariatric surgery, and Sam agreed. This procedure isn't for everyone—there are risks, as with any surgery—but he took them in stride. "No risk, no reward," he shrugs.

For Sam, the results were immediate. "The day after bariatric surgery, I was off of all my diabetes medicine," he says, beaming. Lisa, his wife, who had the same surgery, adds, "It was unbelievable. We left the hospital, and he was told not to take anything . . . the only thing they kept us on was blood pressure medicine." Even better, the weight loss that followed was dramatic. "The heaviest I ever was, was about 323, 324," Sam recalls. "This morning, I weighed myself, and I was 237."

Lisa adds: "I was 241, and now I'm 126."

Most bariatric surgery, including the procedure Sam and Lisa underwent, involves blocking off a portion of the stomach with the theory that the resulting smaller stomach volume will force patients to eat less and lose weight. Some types of bariatric surgery also involve bypassing part of the small intestine, which should diminish the amount of nutrients a patient can absorb from her food. But doctors like Lee Kaplan of Massachusetts General Hospital have begun to suspect that these aren't the only reasons why the surgery works. Studies have suggested that patients who get bariatric surgery often lose much more weight than their peers who didn't get the surgery but eat the same reduced portions of food. So what's going on?

Scientists have shown that the human microbiome changes after bariatric surgery, but does this have anything to do with why the surgery is effective? To find out, Dr. Kaplan and his colleagues performed a specific type of bariatric surgery, called Roux-en-Y gastric bypass, on mice.[11] The scientists first showed that the mice's microbiomes changed composition after the surgery, just as in humans. Good so far. They then transferred the gut bacteria from mice that had the surgery into microbiome-free mice. The mice that got the bacteria from mice who had surgery lost more weight and had less body fat than those that got bacteria from control mice who had been subjected to a mock surgery only or to a mock surgery and a diet. The mouse equivalent of bariatric surgery seemed to be creating a population of gut bacteria that could actually *cause* weight loss. Just as with the malnourished Malawian twins, here again the evidence suggests that the microbiome's composition could play an important role in regulating weight and nutrition.

There are several ways a person's microbial system might become out of whack. The microbial composition she develops early in life could predispose her to certain conditions while protecting her from others. Or the disruption could be a result of substantial and sustained changes in diet or lifestyle that alter the composition of her microbiome. But in some specific cases, human interventions such as antibiotics are to blame.

One downside of antibiotics is that they don't discriminate between so-called good bacteria and bad bacteria. Our microbiome is fairly stable to the stress caused by one course of antibiotics, but when sickly people receive multiple courses over a hospital stay, for example, their gut bacteria can end up in total disarray. When the normal composition of microbes in the gut is substantially disrupted, other bacteria, including some that normally live in the gut in small numbers, can end up taking over. And that can make life miserable, or even prove fatal.

Dr. Don Marcus is a specialist in infectious diseases at a hospital in Pennsylvania. He is one of the doctors who are using a new way of treating one of those bacteria that takes over when its competitors are killed off by antibiotics. This bacterium is called *Clostridium difficile*, better known as *C. diff*, and it kills more than 25,000 people per year in the United States alone.[12] *C. diff* usually takes hold in patients who are on heavy doses of antibiotics, which eliminates its usual competitors and allows *C. diff* to grow out of control. Once established, a *C. diff* infection can cause nasty diarrhea, abdominal pain, a condition called toxic megacolon, and even death in very severe cases.

Treatment can be difficult. *C. diff* is a bacterium, so the usual fix is more antibiotics. As you might imagine, using the same tools that created the problem initially isn't always so effective. *C. diff* gets knocked down for a time, but the disease often recurs, presumably because the bacteria that normally make up the microbiome are also hurt by the antibiotics. When the normal microbes are also reeling, they aren't able to keep *C. diff* from taking over again. These recurrent cases become progressively more difficult to treat. A little fewer than a quarter of *C. diff* cases come back after a first treatment, but the recurrence rate can be as high as 60% for subsequent attempts at treatment. For some cases, it becomes clear that antibiotics simply aren't going to cut it.

Our recent advances in understanding the microbial system in the gut have yielded a treatment that works for many recurrent cases. Since *C. diff* usually takes hold when a patient's system of normal intestinal bacteria is decimated by antibiotics, it makes sense that "re-seeding" her gut microbiome with bacteria from a healthy person might allow those normal bacteria to outcompete the *C. diff*, reset the system, and cure the patient. And the best way to accomplish the re-seeding is with a fecal microbiota transplant.

Calling the procedure a "fecal transplant" is admittedly not the best marketing—some doctors call it a "human probiotic infusion," which is at least a little better—but that's exactly what it is. The doctor will take a small fecal sample from a healthy donor, mix it with a saline solution, and infuse it into the patient's body via an enema or a tube inserted through the nose and snaked down into the gut. Dr. Marcus acknowledges the "tremendous 'yuck' factor," but the bottom line is that patients care about getting better, and the treatment is very effective. Fecal transplants completely cure the *C. diff* infection in anywhere from 80 to 95% of recurrent cases. Dr. Marcus tells stories of patients who were so ill they were candidates for surgery to remove part of the large intestine before this treatment cured them.

For now, a fecal transplant is still a fairly uncommon procedure. The treatment hasn't yet been approved by the FDA—the agency considers feces to be a "drug" for these purposes—and so fecal transplants aren't always covered by insurance.

Fecal transplants are one of the first examples where we've developed our rudimentary understanding of the microbiome enough to produce a novel therapy. Already, work is under way to make the procedure more acceptable to squeamish patients and to improve the consistency of the treatment by developing synthetic feces that can help repopulate the gut. Plus, trials are in progress that will attempt to expand this success to treat other types of diseases that cause an inflamed colon, such as inflammatory bowel diseases (IBD).

One might not expect that the gut microbiome plays a major role in processes and diseases that don't revolve around food. As it turns out, though, scientists have found many cases in which the microbial half of your body-microbiome system contributes to causing diseases or body changes other than weight or gut infections.

Indeed, scientists have long known that both microbes and the gut area generally are constantly talking to the immune system; in some cases, the gut can train the immune system to recognize some things as good and others as bad.[13] For instance, injecting ovalbumin—the protein that makes up most of the white of an egg—under a person's skin normally induces a strong immune reaction. However, this immune response is reduced if the person first eats ovalbumin. Foreign substances ingested by mouth can lead to immune tolerance because the gut tells the rest of the body that they are not a threat.

Some microbes, both viruses and bacteria, can also affect health in a more sinister way: they can contribute to one's risk of developing cancer. Human papillomavirus (HPV) lives in certain types of skin cells, is often spread through sexual contact, and can actually promote cancer development. Cancer is in many ways just a group of cells that divide out of control, and HPV seems to promote this out-of-control growth by affecting the degradation of proteins that guard against cancer-like cell behavior, promoting cell proliferation, and integrating its own DNA into that of the cells it's attacking. The HPV vaccine—brand name Gardasil—really is a modern marvel. Since most cervical cancers are caused by HPV, this is very literally a vaccine that helps prevent people from getting cervical cancer.

HPV is not just a weird outlier, though. Although cancer itself isn't contagious, a substantial fraction of cancers are caused by microbes that we can acquire from our environment. Some scientists estimate that about one fifth of cancer cases worldwide are attributable to known infectious agents such as bacteria and viruses.[14]

One cancer culprit is a bacterium that lives in the gut called *Helicobacter pylori*, which was made famous in 1984 by Barry Marshall, an Australian physician who became convinced that *H. pylori* causes ulcers. Without the means to put together human trials, he experimented on the only person he could: himself. In one of the most don't-try-this-at-home experiments in the history of biology, he drank a culture of *H. pylori* and promptly got *really* sick. When he recovered, though, he had compelling evidence that the bacterium did what he thought it did, and that insight eventually won him half of the 2005 Nobel Prize in Physiology or Medicine.

Subsequent research suggests that *H. pylori* also can lead to gastric cancer. It turns out that the bacterium doesn't grow well in the acidic environment of the stomach, but rather than give up and go home, *H. pylori* produces enzymes that reduce the stomach's acidity by making ammonia. Then the bacterium burrows through the mucosal lining of the stomach to reach a more hospitable environment. These two adaptations disrupt the lining of the stomach, leading to inflammation and ulcers, as Marshall showed. The mechanisms of how it promotes the development of cancer are less clear, but it probably has something to do with *H. pylori* inducing inflammation through biochemical signals or with the potentially cancer-causing chemicals it produces while trying to gain a foothold in the stomach. Either way, *H. pylori* disrupts the system that maintains the stomach's acidity and mucosal lining with sometimes harmful results.

Even though *H. pylori* can contribute to some types of cancer, it might actually prevent other types. That same ammonia production and acidity reduction that can disrupt the lining of the stomach means that you also have less acid reflux, which is linked to certain esophageal cancers. In fact, some scientists hypothesize that the lower rates of *H. pylori* colonization we see in modern people, perhaps because of antibiotics, is the cause of a current sharp increase in esophageal cancers.

Laboratory studies on rats and other model organisms have shown that the microbiome might affect the mental state and behaviors of test subjects. Some studies have even claimed to detect such effects in humans, though the evidence isn't yet credible. We generally like to think that we have full control over our decisions and behavior and that the brain is remarkably well protected from microbial invaders. Could microbes really influence behavior?

One example of a microbe that exerts some influence over its host's behavior is the single-celled parasite *Toxoplasma gondii*—Toxo, for short. Toxo infects warm-blooded mammals all over the world. It can live in a wide variety of hosts, but its favorite host is a cat. The Toxo life cycle begins when a cat ingests Toxo cysts—basically, a hardy and less-active form of the parasite—by eating an infected animal. Once inside the cat, Toxo undergoes sexual reproduction—something that it can do only in cats—and exits in the feline's feces. There it waits until it is accidentally ingested by another host, such as a rodent. In this mouse or rat, the Toxo travels to muscle and brain tissue, undergoes further development, and eventually goes dormant, forming cysts in the tissue. If this infected intermediate host (in this case, the rodent) is eaten by a potential primary host (a cat), the Toxo infects the new feline and the cycle begins again.

Although Toxo reproduces sexually only inside cats, the microbe can reproduce and spread asexually in most animals. Sexual reproduction is likely advantageous for Toxo for the same reasons it is for other organisms: sex allows an organism to mix up its genes with the genes of a partner, creating diverse offspring and new combinations of genes; this diversity allows these offspring to thrive in changing conditions and under new evolutionary pressures. Consequently, it is highly beneficial for Toxo to end up in a cat, and for its progeny to find their way back inside another cat.

When Toxo finds itself inside a rat or mouse, it has a problem. Millions of years of evolution have programmed these rodents to avoid predators such as cats,[15] but Toxo needs its host to be eaten. So what's an enterprising parasite to do? To an outside observer, it looks like mind control. Through some as-yet-unknown mechanism, Toxo seems to make its rodent host unafraid of cats, or at least the smell of them.[16]

To investigate Toxo's effects on rodents, four scientists at the University of California, Berkeley, used common laboratory mice.[17] When the Berkeley researchers took normal lab mice and put them in cages, the mice wandered around the space that was available to them. When the scientists put a dish filled with bobcat urine at one end of the cage, the mice spent significantly less time next to it and more time in the other end of the cage, doing their best to avoid the scent of a predator, insofar as that was possible within their enclosure. When the dish was replaced by rabbit urine, however, the mice behaved as if there were no urine there at all.

The researchers then took other mice, infected them with Toxo, and subjected them to the same tests. Amazingly, the infected mice—unlike normal mice—showed no aversion to their predator's urine! The scientists ruled out general olfactory malfunction by making sure that the mice could find food by smell just as well as normal mice do, so the Toxo infection seems to be causing these mice to change a deeply ingrained predator avoidance behavior. Additionally, when the scientists infected the mice with a strain of Toxo that tends to leave the body over time, the effect persisted even after the parasite was no longer present at detectable levels in the mouse's brain tissue. This observation would suggest that the parasite causes some change in or damage to the mouse that lingers even after the infection has cleared—perhaps an injury to a critical part of the brain.

The jury's still out on what precisely Toxo is doing to these mice. Whether it is selectively destroying a mouse's ability to smell cat urine or doing something even more primal with the rodent's "fear" response,[18] Toxo seems to have developed a fascinating method of integrating itself into the mouse's body and altering it in a way that drastically changes the behavior of its host. And recent evidence suggests Toxo does something similar in chimpanzees, where it seems to render them unafraid of leopard urine.[19]

Toxo, which can infect most warm-blooded mammals, infects many humans. About 10 to 20% of the United States population is infected with Toxo, and the rates are even higher in the developing world. Overall, about 2.5 billion people worldwide are infected with Toxo . . . and for most people, that's probably not a big deal.

Toxo can definitely be dangerous to highly immunocompromised people; the parasite can cause symptoms ranging from headaches and nausea to seizures. Toxo can also be a problem for pregnant mothers; the parasite is normally in a dormant state in humans, but pregnant women with an active form may risk passing it to their developing fetuses, which can trigger complications. In most healthy people, however, long-term infection with Toxo has long been considered asymptomatic. Some studies have found a correlation between long-term Toxo infection and schizophrenia, slowed reaction times, and even suicide. Most of these studies are attention-grabbing but ultimately unconvincing; they tend to conclude that people with Toxo infections are only very slightly more likely to experience the problem under study, and many of the studies have methodological shortcomings that prevent them from showing that Toxo is actually causing the behavior. For now, most people should worry more about their cat allergy than about getting Toxo from those cats.

Weird parasites aside, the connections between the body and the normal, non-parasitic microbes that live in it are so extensive that it's plausible they could affect brain functions such as our mood. For example, many bacteria that live in the gut produce the same types of chemicals that our neurons use to communicate with one another, and those chemicals could be sending signals to the body.

In non-human animals, studies show that gut microbes can influence behavior; for example, scientists have demonstrated that a particular strain of bacteria can affect anxiety in laboratory mice. Anxious mice tend to be skittish when traversing high places or open spaces, and they have higher levels of stress hormones and certain types of signaling in the brain. When scientists fed some of these mice a bacterium called *Lactobacillus rhamnosus*, they relaxed dramatically on all of these measures,[20] indicating that merely having this bacterium in residence in the gut could actually alter the mouse's mood.

This effect seems to be dependent on communication between the stomach and brain. When the scientists cut the vagus nerve, the main communications link between the stomach and the brain, the effect disappeared. That means that whatever effect *L. rhamnosus* is having on the mice depends critically on that nerve's operation—which gives new meaning to thinking with one's stomach.

Other groups have found similar results with other types of bacteria,[21] and gut bacteria can affect other processes in the central nervous system such as inflammation.[22] There's no good direct evidence for such effects in humans yet, but more research is needed.

CHAPTER THIRTEEN

This Is Your System on Drugs

Tweaking Biological Systems to Produce Better Medical Treatments

When hiking at an altitude of 4,500 meters (15,000 feet), the body is working with about 40% less oxygen in the air than is available at sea level. Some travelers who go to Peru looking for mountains first find an annoying headache; others can be thrown for a loop for a few days with symptoms such as nausea. There are modern medicines to help travelers who fly from sea level to high altitude cope with the change, but most of them aren't actually all that effective. The only thing that works for sure is spending time acclimating to the thin air.

Native Peruvians have been dealing with altitude for a long time, and they have a solution that they swear by: the coca plant. They stuff coca leaves in their cheeks—they don't chew the leaves, just suck on the wad like a lollipop—and they drink coca tea pretty much constantly. Coca leaves contain a mild stimulant that feels similar to caffeine in its effects, and Peruvian guides are adamant that coca is the solution to any altitude-related problem that might arise. The tea tastes all right, but on my recent trip, I tended to avoid it after my first try. Like caffeine, it made my heart race.

The active ingredient in coca tea is small amounts of various alkaloids, including cocaine.[1] (As you can imagine, it's illegal to bring the tea back to the United States.) But what's especially interesting about coca is how this drug* was developed. The Quechua people who made up the Incan empire and live in modern-day Peru have been using coca as

* The word "drug" can mean a lot of different things; to scientists, a drug is simply a chemical that has an effect on the body. Both cocaine and caffeine are drugs.

a remedy for thousands of years. It's unlikely that the ancient Quechua people decided that they needed a drug to combat altitude sickness and embarked on a big search for a suitable candidate. Instead, they explored the pharmacological properties of many plants around them to discover remedies for what ailed them.

Finding cures for Alzheimer's disease and HIV will require more complex processes than informal experimentation. The modern Western pharmaceutical industry has to do things very differently. There have of course been instances of drugs being discovered accidentally in a formal research setting—penicillin, for example, resulted from a chance observation that mold growing in the lab was killing the bacteria that Scottish biologist Alexander Fleming intended to do experiments with—but pharmaceutical research and development can't simply rely on good fortune. Instead, there is a relatively structured process by which we develop drugs.

Typically, pharmaceutical drug design progresses by focusing on a particular target. Let's say we know of a protein, maybe a receptor that sits on the outside of the cell, that we might be able to turn off or on in order to affect a certain disease. For example, I'm a fidgeter, so I'm always bouncing my left leg without thinking about it. Let's imagine that there's a certain protein in all of the cells in my leg that, when activated, sends a signal that keeps me from fidgeting. In most people, there's a chemical floating around in the bloodstream that turns on this anti-fidgeting protein, but I don't have enough of that chemical.

In this case, one path to a drug is pretty clear: develop a chemical that binds to the anti-fidgeting protein and turns it on. To do so, modern pharmaceutical companies often try hundreds of thousands or even millions of compounds in very simple model systems. That way, they can easily test many candidates in the lab in order to see which are worthy of further investigation. Researchers take the compounds that perform well in this initial screen and explore variations on them before ultimately testing them in animals, then humans. At the end of the process, they hope to find a drug that is both safe and effective.

The only problem is that this process doesn't seem to be working very well. There's been a lot of talk in the pharmaceutical industry for a decade or two that something's not quite right. The first clue is that the costs of developing a new drug are enormous and growing. The exact number depends on how you count, but drug companies spend something like two billion dollars on research and development for every drug that eventually gets approved.[2] One big contributor to these costs is the fact that drug discovery is hard; the vast majority of candidates that are tried either don't work or aren't safe. And despite all of the research and testing that we do, most drugs that fail human trials fail because they simply don't work. There seems to be some promise in a new wave of approved "biologics," or medicines made from proteins that have been engineered to treat a disease, but there's clearly still room for improvement.

There are probably many reasons why the pharmaceutical industry isn't working well, and everyone has different ideas about how to fix it. Some say that we've already picked the low-hanging fruit and are stuck designing drugs that are somehow more difficult to make than the ones that came before. Others say we have a misguided focus on developing small molecules that we can synthesize chemically instead of searching for chemicals from plants or other natural products that can do the job. Still others point to a myopic focus on rules that purport to say what chemicals are "drug-like," or to the trend for our drug candidates to get larger as we optimize them, or the fact that drug companies are currently incentivized to close down research and development to focus on short-term profits.

Another possibility is that sometimes we simply don't understand how the target behaves in an actual human being. Some drugs fail because we don't understand the biology of the system well enough, or how the specific component that we're targeting works within the overall system. In these instances, the drug wouldn't fix the disease even if it worked exactly as designed. After all, fixing a biological system with a drug that specifically blocks a single protein or component of a pathway is kind of like trying to fix a car engine by shooting it with a sniper rifle. While there are certainly problems that can be fixed this way, having a little more control could lead to better outcomes.

Considering the full dynamics of the relevant biological system when designing a drug would definitely be the ideal. But the systems approach seems quite difficult to apply in practice: there are very few diseases that we really understand on the systems level in humans, and we don't typically want to wait until science has figured everything out before trying to make a drug.

Until our understanding of the systems that underlie disease improves, one possible strategy is to use drugs that take advantage of a small, local system in the target protein itself. Imagine that there's a disease that's caused by scissors cutting too much paper. The typical approach might be to jam the scissors with a barrier that the scissors can't cut through:

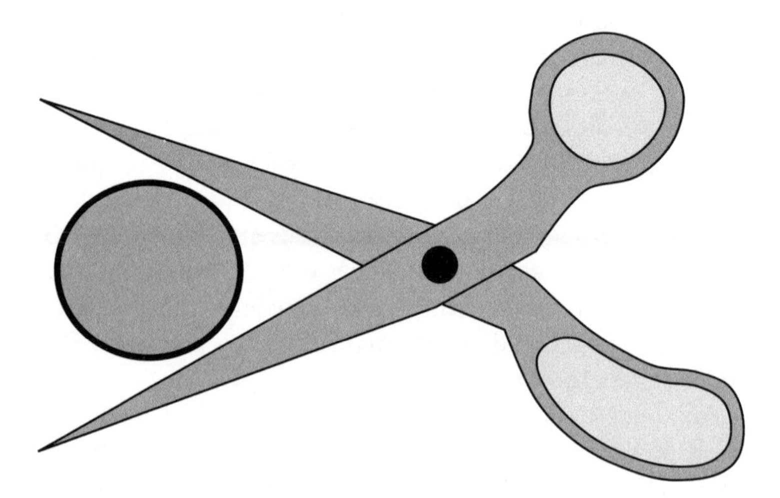

Figure 55: A jammed pair of scissors.

Another option is based on knowing that the blades are connected to the handle. For instance, we could put a rubber band around the handle that would make it hard to open the blades and cut anything:

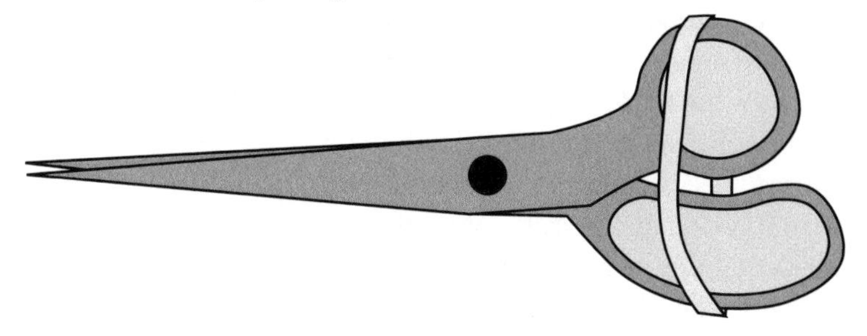

Figure 56: A pair of scissors restrained with a rubber band.

In this case, we didn't have to shove something in the place where the cutting happens, we just pushed on another part of the scissors. When looking at a protein, similar principles can hold. We might be able to design a drug that will bind on one part of the protein and have an effect somewhere else. These kinds of drugs are called allosteric drugs. The first part of the word, *allo*-, comes from the Greek word for "other,"[3] so allosteric refers to a drug that exerts its effect by binding or grabbing on to the protein in the "other" site.

There are a lot of reasons that drugs that work like this might be beneficial, but most of them boil down to the potential for making more effective drugs with fewer side effects. For example, let's go back to the hypothetical anti-fidgeting drug. If I don't have enough of the natural signaling chemical that activates the anti-fidgeting protein in my legs, the traditional approach might be to try to design a drug to turn on the anti-fidgeting protein. Because of the way we do the design process, we'll likely end up with a drug that grabs on to the anti-fidgeting receptor and turns it on in a way that's very similar to the way the natural chemical in my body does. Then we can use more and more of that chemical to turn the protein more and more "on." But in this hypothetical, that protein would turn on in pretty much every cell in the body. Maybe that's the right thing to do in the legs, but perhaps turning on that protein in the heart causes an undesirable side effect.

What if we instead designed an allosteric drug that could bind somewhere else on the protein and make the anti-fidgeting chemical signal work more strongly? In that case, we might not have the same problems with side effects. Instead of turning on all of the anti-fidgeting proteins in all of the parts of the body all the time, we'd just make the signal stronger in places where it already is, like my leg. In some sense, this allows us to fine-tune existing biological responses, which could cause fewer accidental problems. Even better, if the amount of the natural signaling chemical ordinarily changes with time—say, there's a lot of it, then a little, then a lot again—allosteric drugs can preserve that pattern, which might make the drug more effective or allow it to have fewer side effects.

Allosteric drugs might also be able to hit their target more specifically than "normal" drugs. To continue the fidgeting example, imagine that there are actually five different proteins that look very similar: an anti-fidgeting protein, a painful-leg-cramps protein, a muscle-weakness protein, a temporary-paralysis protein, and a turns-your-skin-orange protein. Obviously, the ideal would be to turn on the anti-fidgeting protein while leaving the others untouched, but all of these proteins are almost identical in the part where the anti-fidgeting chemical binds. It's essentially impossible to design a drug that works like the natural chemical signal that doesn't accidentally turn on the other four undesirable proteins. But if we could design a drug that binds somewhere on the anti-fidgeting protein that *is* different from the other four, we would be able to avoid all of those undesirable side effects.

Allostery in a Different Age

In the 1904 Olympics marathon, Thomas Hicks employed a strategy that sounds odd to modern ears. Over the course of the race, his assistants gave him egg whites, brandy, and a few milligrams of strychnine to perk him up whenever he started to fall behind. Reinvigorated by these treatments, he performed well: Hicks was the second person to cross the finish line. He was eventually declared the winner when race officials discovered that the first-place finisher had driven about 11 miles of the race in a car.[4] It was a different time.

One of Hicks's remedies, strychnine, is most commonly known as a deadly poison. At low doses, though, it can act as an allosteric, performance-enhancing drug: it interferes with the body's signaling and makes muscles respond more easily. It was relatively popular in the early 1900s; indeed, a character in H. G. Wells's *The Invisible Man* sings its praises: "Strychnine is a grand tonic . . . to take the flabbiness out of a man."[5] Perhaps unsurprisingly, it has since fallen out of favor.

Scientists and doctors have tried to understand allosteric drugs for many years, but they haven't always known exactly what they were doing. In this category of "early exploration" is a study done by four doctors in 1947. Some surgeons had taken to using a new drug, curare, as anesthesia during surgery, but no one was totally sure whether it made patients numb or simply paralyzed them. This was a potential problem because curare was sometimes used as the only anesthetic during surgery, "especially in infants and children." The prospect of operating on a paralyzed infant who could still feel everything was apparently appalling even in 1947, so these doctors decided to figure out if curare actually did numb pain.

One of the doctors volunteered to take the drug, and the group developed an elaborate series of body language signals to let the volunteer report what he was feeling. They gave him higher and higher doses of the drug and kept asking him if he could still feel them prodding and pinching him. By the end of the experiment, the volunteer doctor was lying on a table, almost totally paralyzed, able to communicate only by blinking—and reporting that he could feel pain just fine, thanks very much. The study casually mentioned that upon regaining movement, the volunteer said that being unable to swallow had left him feeling like he was drowning for a full twenty minutes!

Because of the potential benefits of allosteric drugs, pharmaceutical companies are interested in searching for them specifically. Though new screening methods have made finding allosteric drugs easier than it once was, the process of finding an allosteric drug is still difficult—more challenging than it is for traditional drugs. New tools would come in handy.

One such tool sits on the 32nd floor of a nondescript office building one block off Times Square in New York City. There, inside a glass-walled "fishbowl," sits a very specialized computer about half the size of a small moving van. It can't play solitaire, check e-mail, or run Windows. Indeed, this computer was built for one purpose only: to simulate the way proteins change shape and interact with other molecules in our bodies so that we might one day be able to design better drugs.

The man behind the computer is David E. Shaw. Shaw began his career as a computer science professor at Columbia University, but he was lured to Morgan Stanley in 1986 to help develop their computing infrastructure for analyzing financial markets. Soon after, he left to start his own hedge fund, D. E. Shaw & Co., which was one of the first and most successful firms to use computer algorithms to drive trading decisions.

But after nearly 15 years in finance, Shaw gave up the management of the hedge fund that bears his name and turned to a new, audacious project. Conversations with his friends who worked in biology and chemistry had convinced Shaw that the proper application of computing power could stimulate significant progress in those fields, much as it had changed Wall Street. Specifically, Shaw saw an opportunity to build a computer that could help scientists discover new pharmaceutical drugs—so he assembled a team and went to work.

Shaw and his group—D. E. Shaw Research—spent their first six years designing a special supercomputer that can very efficiently simulate how drugs and proteins work. People had been using the simulation technique that Shaw and his team settled on for a long time, but the method had been limited by the vast amount of computing power it required. This limitation made it very difficult to run simulations long enough to observe many biologically and pharmaceutically important processes. Shaw's group solved this problem by tailoring the circuitry of their new machine to the types of calculations needed for these sorts of simulations. They called the resulting supercomputer "Anton," after Antonie van Leeuwenhoek; a Dutch scientist at the turn of the eighteenth century, Leeuwenhoek did pioneering work in improving and perfecting the microscope. Shaw pictured Anton as sort of a "computational microscope" that would let scientists watch simulated proteins and drugs interact, as if through the world's smallest video camera.

They eventually built more than a dozen Antons. One stayed in the office in the fishbowl, one was lent to the Pittsburgh Supercomputing Center for use by other scientists, and the rest were installed in a warehouse in upstate New York; Shaw was told there wasn't enough room under the streets of Manhattan to run the electrical cabling the full set of computers would have required. And with Anton up and running, Shaw and his team began using it to study biology and to research new potential drugs.

The simulation technique that Anton uses, molecular dynamics, is actually rather simple at its core. In a typical application, Anton might simulate a biological system consisting of a drug, the protein it targets, the cell membrane the protein resides in, and even the water that surrounds it. All of the atoms in the simulation are represented as tiny balls.[6] Anton then uses a set of rules—a "force field"—that describes the interactions among all of the atoms.

Amazingly, the simulation's rules aren't that much more complicated than the physics that students learn in high school; Anton essentially takes Newton's laws of motion and applies them over and over again. The simplest rule governs bonds between two atoms: these bonds behave like springs and tend to keep two atoms at a given distance apart:

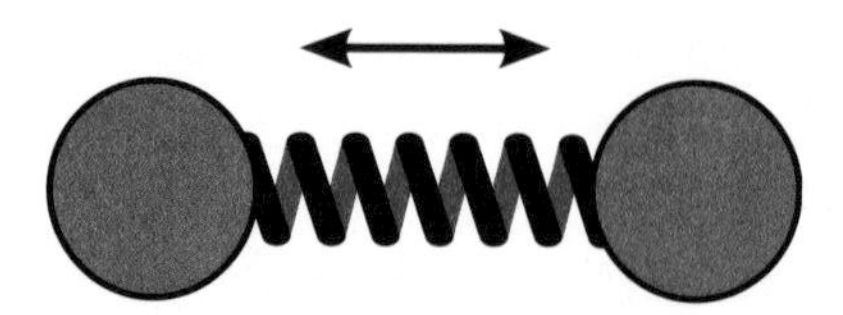

Figure 57: A bond between atoms, represented by a spring.

Other rules govern the angle between two bonds:

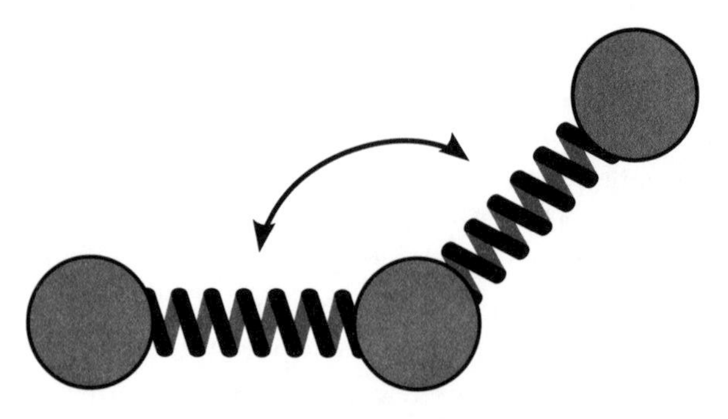

Figure 58: An angle between two bonds.

There are also a few rules that put restrictions on the relative positions of four atoms. But that's about it: these simple rules, combined with calculations of the electrical interactions of charged atoms, are the core of Anton's simulations.[7]

But what Anton lacks in complicated physics it more than makes up for in scale. A typical system might easily contain several hundreds of thousands of atoms. That means that

Anton must evaluate tens of billions of particle-particle interactions every time it wants to advance the simulation by a single time step, which is usually around 2 femtoseconds (for reference, a femtosecond is about 100 trillion times shorter than the blink of an eye). To simulate the behavior of a biological system for just one thousandth of a second, Anton has to apply its rules over and over for about a half trillion time steps—that's like simulating the retreat of the glaciers from New York State by checking their progress every two seconds over thirty thousand years.

After that enormous amount of computation, scientists are left with a large amount of data describing where every atom is throughout the simulation. Up close, it looks like just a bunch of balls and sticks:

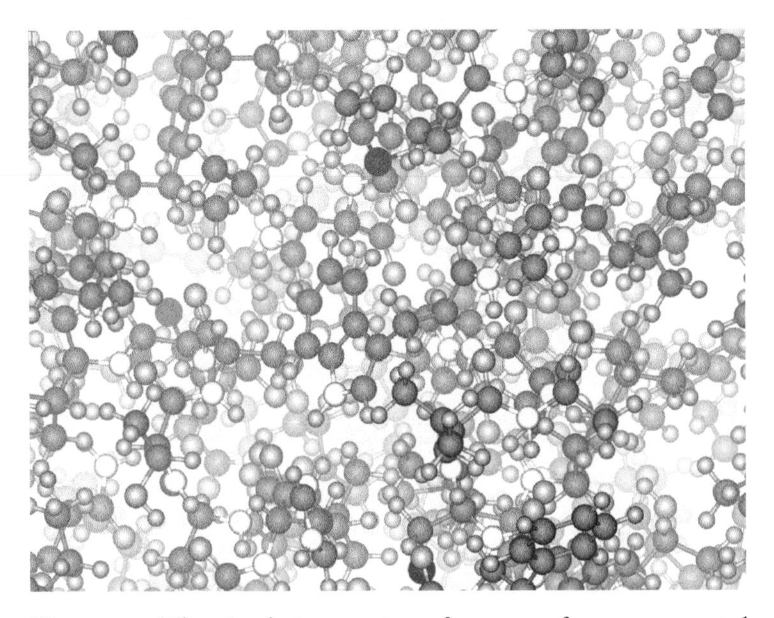

Figure 59: The simulation consists of a soup of many connected particles.

We can draw this in more human-friendly form by zooming out, leaving out some of the atoms, and drawing others with cartoons. Below, the protein is drawn as a ribbon, and it sits in the cell membrane—the barrier that encloses the cell—which is drawn using a stick-like representation. All of the remaining space is taken up by water, but for clarity, those water molecules aren't shown. The result is much more understandable:

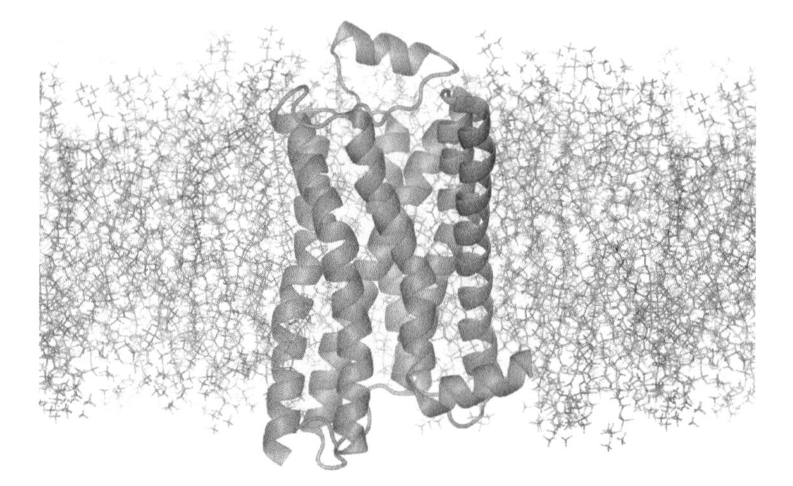

Figure 60: Drawing the protein as a cartoon helps make the simulation results easier to interpret.

With this representation, scientists can watch the simulation unfold like a movie. In Anton's simulations, proteins move realistically, and drugs float around and bind to the protein spontaneously. This allows researchers to gain insights into how the drug and protein interact that few other techniques can match.

Of course, as with any model, you can't blindly believe what you see; molecular dynamics simulations do have major limitations. Those rules that describe how the atoms interact can't capture all of the details of how real molecules behave. For instance, these simulations cannot account for all of the effects of quantum mechanics, the physics theory that deals with things that are very small. Luckily, the rules that Anton uses turn out to be good enough to get some insights into biological phenomena. Using a combination of simulation and careful validation by experiment, scientists can get a good idea of how a drug is actually behaving.

A team of scientists from D. E. Shaw Research published a paper in 2011 that first drew my attention to the use of molecular dynamics for drug discovery.[8] In that paper, the researchers investigated drugs known as beta blockers, which are commonly used to treat heart conditions. Beta blockers get their name because they bind to and inhibit the activity of proteins called beta-adrenergic receptors. In those simulations, the researchers allowed the beta blockers to float around the simulation freely, and the drugs spontaneously bound to the receptors. Using this information, the group described in detail the mechanism by which the beta blockers bind.

These simulations were impressive for two reasons: first, the researchers had been able to validate their simulations by showing that the final position of the drug on the protein in the simulation matches its known position in reality essentially perfectly. And second, the simulations showed in exquisite detail how the drug goes about latching on to its target receptor—a result that would be difficult or impossible to get with current laboratory equipment.

After I graduated from college, I moved to New York for two years to work with Shaw's research group before I started graduate school. One project that I worked on, led by Dr. Ron Dror, now an associate professor of computer science and molecular and cellular physiology at Stanford University, focused on allosteric drugs that bind to G-protein-coupled receptors (GPCRs). GPCRs are a type of protein that sits in the cell membrane, which separates the inside of the cell from the outside environment. The particular GPCR we studied, called M2, looks a bit like a sock puppet with its mouth poking out of the membrane (see Figure 61).

When signaling molecules from outside of the cell bind to M2, it changes shape, and that shape change carries a message to the inside of the cell (see Figure 62).

Depending on the GPCR, that message could tell the cell to grow, to fire an electrical impulse, or to commit suicide.

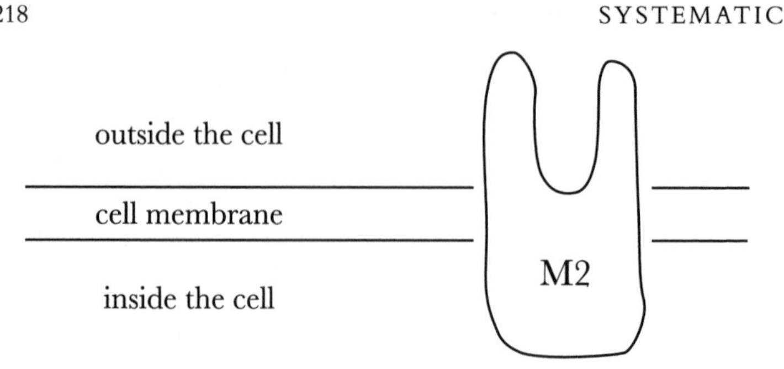

Figure 61: The M2 receptor.

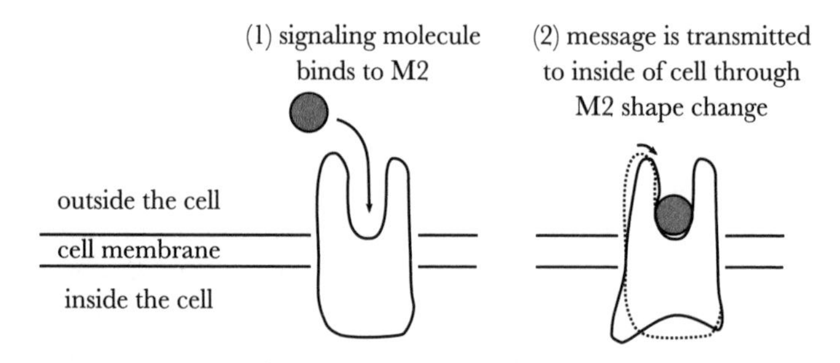

Figure 62: M2 changes its shape when the signaling molecule binds to it.

M2, for instance, helps control one's heart rate, among other functions. There are about a thousand different GPCRs, and approximately one third of all pharmaceutical drugs target them,[9] so understanding how GPCRs work is of great interest to drug companies.

When we started to work on M2, scientists knew of a few allosteric drugs and other molecules, such as strychnine, that grab on to M2 and alter the signals it sends. We knew that these allosteric drugs could bind somewhere on the protein and make it either easier or more difficult for the normal signal to exert its effect on the receptor. No one knew exactly where these allosteric drugs bound to M2 or how they affected the signals that pass through it, however. We wanted to find out.

We started off by letting the drugs float around freely in our simulation, and they spontaneously grabbed on to M2. Each drug has a few charged regions, and we saw that these regions were helping the drug stick to certain areas of M2, like a magnet sticks to a refrigerator. All of the allosteric drugs we looked at seemed to latch on to the top of the crease in M2:

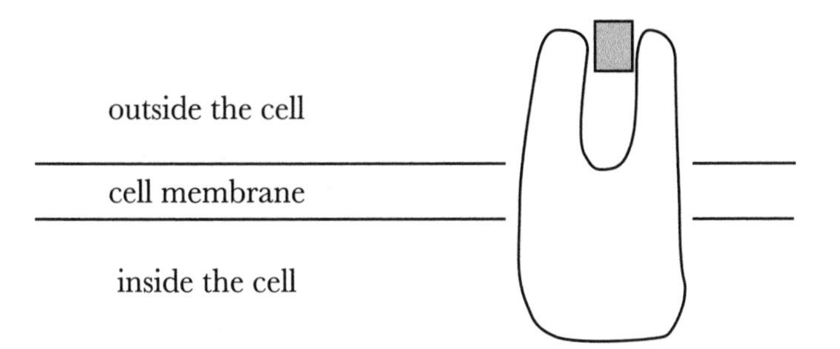

Figure 63: The allosteric signals seemed to bind at the top of the crease.

The normal signaling molecules from the body, in contrast, tended to bind deeply in the crease:

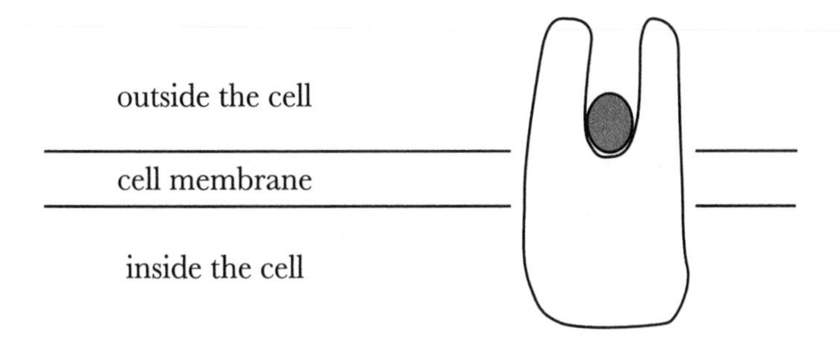

Figure 64: The usual signals bound lower down in the crease.

This deeper pocket is considered to be the primary or "orthosteric" binding pocket.

The challenge then became understanding how these allosteric drugs communicate with the usual, orthosteric signaling molecules. One way became immediately obvious: since both the allosteric drug and the orthosteric signaling molecules are positively charged, they tend to repel one another. This mechanism would predict that having the allosteric drug bound to the receptor would make it harder for the natural signal to bind, thereby reducing the strength of the signal. By doing some simple calculations, we confirmed that this effect should indeed be important for allowing an allosteric drug to impact signaling.

But we also knew that this could not be the whole story. Some allosteric drugs make the normal signals *more* likely to bind, which can't be explained by a model where these molecules simply repel one another. Something else was happening.

The key insight came from another observation: some of these allosteric drugs had to force the crease of M2 to open wider than others:

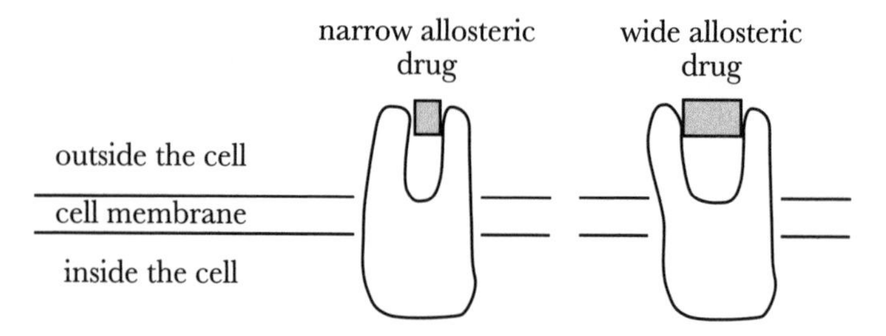

Figure 65: Different drugs forced M2 to adopt different widths.

This pointed to one major way that these allosteric drugs can affect the normal signals. Some of those normal signaling molecules also need a wide crease in order to bind to M2. A

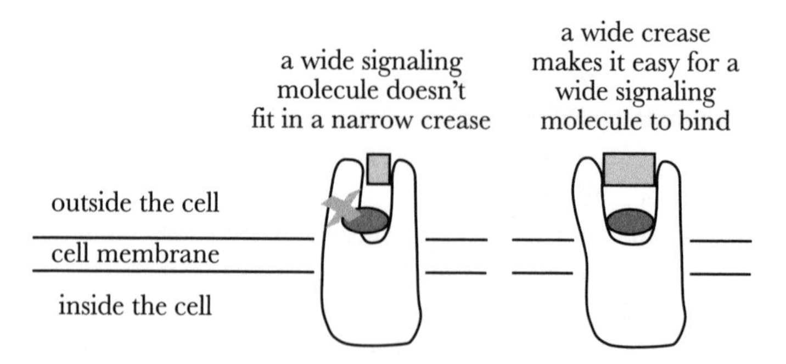

Figure 66: Wide allosteric drugs could make it easier for wide orthosteric molecules to bind.

"wide" allosteric drug has already done the work of opening up the crease, making it easier for a wide signaling molecule to bind and send a signal.

Once we knew that the width of the crease was important, we tried using that knowledge to modify an existing allosteric drug. We reasoned that making the drug "wider" should cause it to favor wide signaling molecules. Indeed, when we designed a wider version of the drug, that's exactly what happened.

Showing that we could intelligently modify an allosteric drug was exciting because improving a drug is usually a trial-and-error process. Chemists will often make hundreds or even thousands of versions of a candidate drug to try to make it better. That process, which drug researchers call exploring the "structure-activity relationship" (SAR), is time-consuming and often doesn't yield much insight. A well-developed SAR might tell the researchers that versions of the drug that carry an electrical charge at a given position tend to bind better to the target. But it won't tell them why that works, or what other options might be better.

By contrast, using a simulation to help guide drug discovery could make the process much more reasoned. Instead of blindly trying many different variations of a drug to see what works, scientists with a full understanding of how the drug grabs on to its target could make considered changes: "We should put a charged group at position 5 because it will help the drug attach to the lysine at the 132nd position in the protein." This is often called "rational" drug design.

One of the most exciting aspects of our work was that we were able to use this rational approach to change the allosteric properties of a real drug: we figured out how the drug works and tweaked it using that knowledge. If that process can be applied to other drugs, it could make the drug discovery endeavor move much more smoothly.

Alongside computational approaches to developing new drugs, there are many other new technologies hoping to become an invaluable part of the drug discovery process. One of those technologies aims to improve the safety of the drugs we do produce.

The scientists who develop drugs spend a lot of time, energy, and money to make sure that the compounds they develop won't harm the people they're trying to help. Any drug that is approved for use in humans has to meet very stringent standards to make sure it's safe to take. There are a few exceptions. If a late-stage cancer patient likely has only a short time to live, we'll consider drugs that have some risks or nasty side effects as long as they have a decent chance of preventing the cancer from killing. But we don't accept those same sorts of risks in people who are mostly healthy. One has to be quite sure that drugs like blood pressure medications, for example, aren't going to harm the person taking them.

That's not to say the pharmaceutical industry has never made mistakes, though. One infamous example is Vioxx, a prescription painkiller that was commonly used for arthritis pain in the early 2000s. It was pulled from the market when it became clear that it increased the risk of a heart attack under certain circumstances. That safety failure led to some changes in the industry: it's now common to run drug candidates through a test that makes sure they don't interfere with the heart in the way Vioxx did. But the vast majority of drugs are either perfectly safe or beneficial enough that they're worth some risks.

Making drugs safe is pretty difficult. Part of the problem is that scientists have to develop drugs in test tubes and petri dishes and mice. As useful as experiments in those kinds of systems can be, none of them are human, and in many cases, they don't behave all that similarly to human systems. If we could make our test systems behave more like human systems, we might be able to improve our ability to estimate how a candidate drug will work in a human patient.

One person who's trying to solve this problem is Vyas Ramanan, a graduate student at MIT. He cofounded a company that makes better devices for performing medical procedures in small blood vessels, and in his spare time, he worked with a team that is trying to use drones to deliver vaccines to the developing world. His profile on his lab's website promises that their plan "does not involve actually vaccinating patients from a drone . . . yet."

Ramanan works on, among other things, building better models of human livers for use in drug testing. The liver is responsible in part for breaking down chemicals in the blood, and most drugs are carried around in the bloodstream. "You know that it's going to pass through the liver because basically everything does if it's in your bloodstream for even a short amount of time," says Ramanan. The liver then has a chance to break down a drug, and we might not always like the result. Even if the original drug isn't toxic, Ramanan says, the chemicals that the liver breaks the drug down into might be. Scientists have to worry not only about making the initial drug safe, but also about the possible effects of any of the by-products that are produced as the drug is degraded.

Perhaps the most famous and tragic case of a toxic by-product was an early-1990s anti-hepatitis B drug called fialuridine.[10] None of the testing in various animals indicated that the drug might be toxic, so it was moved into human trials. The initial trials lasted only two to four weeks, and the drug didn't seem to be producing any ill effects. The hepatitis B virus didn't seem to be responding quite as well as the doctors had hoped, though, so in March 1993 they started another trial in which fifteen patients would receive the medicine for a longer period of time. After eight weeks, things were looking great—the hepatitis B seemed to be responding to the treatment and the patients were reporting no major side effects. In the next few weeks, a few patients started to complain of symptoms like fatigue and nausea, but those aren't really terribly out of the ordinary. Suddenly, in week 13, one of the patients was admitted to the hospital with liver failure. That's a really big problem, and it called for emergency action.

The doctors immediately stopped the trial and brought everyone in for examination. They found that seven of the patients had varying degrees of liver damage. Despite the doctors' best efforts, those seven patients continued to deteriorate over the course of several weeks. Out of the original fifteen patients, five died and two more survived only with liver transplants.

As tragic as the deaths were, it's not obvious that the scientists and doctors who were developing the drug should have seen this coming. None of the animals that they tested the drug on before giving it to humans showed any warning signs. So what could we do to prevent a tragedy like this from happening in the future?

One possible solution is what some scientists call "organs on chips." An organ on a chip is a little plastic shell that contains human cells—liver, heart, brain, whatever—that are intended to mimic the full human organ. We can then use those modules to test a drug for toxicity. "Let's say you have a liver module hooked up to a heart or a brain module," Ramanan says. You would send the drug through the liver module and observe the drug's effects on your liver model. Then the outflow from the liver module—which contains the drug in whatever broken-down form it might have after passing through the liver—might flow into the heart or brain module. "You can then look at the effects of these converted drugs on the heart cells and the brain cells," Ramanan says, "and you might see vastly different effects" from what you would have seen with the original drug. Basically, we would be reproducing the relevant parts (in this case, organs) of the human system and testing the drug on that model before putting people at risk.

There's an additional complication: not all people have the same liver, the same heart, or any other component that a drug might impact. Some patients have alterations—maybe a mutated component that has a different level of activity—that change how the system behaves. And since each of these variations is different from patient to patient, perhaps the best thing to do would be to change the treatment depending on the individual.

With all we've learned about cellular systems in the past decades, we've sometimes been able to tailor our treatments for certain diseases based on our knowledge of what's wrong with that patient on a molecular level. For example, we can use sequencing technology to look at changes in important genes and try to understand how those changes might cause the disease. And in some cases, we've been able to take a group of patients with a disease, figure out which of them will respond to a given treatment based on the defect that caused their disease, and treat each of the patients with the drug that will be most effective for them.

As discussed earlier, genomic sequencing has allowed us to figure out that the cancers we call "breast cancer" are actually multiple types of cancer that all happen to develop in the breast—and that those different breast cancer subtypes all respond differently to treatment. That's one example of tailoring the treatment to the individual's genetic makeup, but there are other examples in cancer where we have actually seen a specific defect in some patients and designed a treatment for it. One of the best-known examples of this is the cancer drug imatinib, more commonly known by its brand name, Gleevec. Scientists discovered that some types of leukemia have an alteration in their DNA that creates a weird fusion of two proteins that helps make that cell cancerous. Gleevec was designed to treat this specific defect by turning off the defective protein, which can halt the growth of the tumor cells or kill them. This treatment radically improved survival in patients with that kind of cancer.[11]

We've had a lot of success with this strategy in cancer, but other diseases have been a bit harder to crack. One possible reason for this is that we don't really have the ability to personalize treatments on the level of systems yet. Most of the examples of personalized medicine are "local," in the sense that we see one component of a system that is altered in the same way in many different people with the disease, and we try to fix that one component—much like the example of the Beery twins discussed in Chapter 5.

But some diseases, like autism, asthma, schizophrenia, and multiple sclerosis, are harder to ascribe to just one or two molecular defects that occur over and over. Some scientists think that diseases like these arise partially from one or more rare variants—that is, many instances of the disease have slightly different and usually novel causes. There are probably a lot of ways for any given system to break, so one frontier for a systems approach to medicine is predicting how any given defect will affect the system in question—and how we can use drugs to fix that system.

One other exciting idea is to test a drug on an artificial system made out of the patient's own cells.[12] Since we can now take a person's skin cells and turn them into any other kind of cell, as discussed in Chapter 5, we might be able to actually test treatments on your cells before giving them to you—or to use those cells to help figure out what is going wrong.

But while all of these other possibilities are still developing, we've already had a lot of success with cancer—and we're doing a lot better than we used to. In his fantastic piece "A Note from History: Landmarks in History of Cancer, Part 2" published in the journal *Cancer* in 2011, Dr. Steven Hajdu describes how we've thought about and treated cancer in the past. In the "swing and a miss" category: some seventeenth-century doctors thought that breast cancer was contagious.[13] But in some other ways, those doctors were actually pretty advanced; for example, the German physician Johannes Scultetus described performing mastectomies in a 1655 book published after his death.

These days, we know a lot more about cancer. It is a really difficult disease to treat because cancer is actually hundreds of different diseases. As mentioned previously, a tumor is actually just a group of cells that keep on growing when they shouldn't, and there are a lot of ways for that to happen. The body has many defense mechanisms set up to curb the growth of runaway cells, but when those mechanisms fail, the results are devastating. Cancer starts in one location and can invade others, disrupting vital organs and ultimately killing the patient.

Beating cancer is about more than just treating the tumor itself. In fact, the tumor and its environment interact in complicated ways that make it important to think about the tumor and the surrounding tissue as a connected system. In our early screens for cancer drugs, we often looked at cancer cells by themselves. Unfortunately, cancer cells behave rather differently in their "natural" environment from when they are alone in a dish. In fact, sometimes the surrounding cells actually help to protect the cancer cells from our drugs.[14] But luckily, we can use this knowledge to figure out what those interactions within the tumor-environment system are and to block the ones that are helping the tumor survive.

One of the most difficult parts of treating cancer is that cancer cells are really just your own cells that are misbehaving. That makes it really hard to kill cancer cells and only cancer cells. Traditional chemotherapy targets fast-growing cells, which includes not only the cancer but also all sorts of other important cells, which leads to the horrible side effects chemo patients must endure. For example, these fast-growing cells include the cells that make up the hair follicles, so patients undergoing chemotherapy often lose their hair. Targeting cells that are fast-growing in an attempt to kill cancer is like trying to use a sledgehammer to crush one specific ant in a group: sure, it can get rid of the target ant, but it will take out a lot of innocent ants in the process. A really successful cancer treatment would attack cancer cells and only cancer cells.

The problem of minimizing collateral damage during cancer treatment in some ways resembles trying to keep spam from overwhelming your e-mail inbox. When classifying spam, we want to delete all of the spam e-mails but keep all of the legitimate e-mails. To do this, we can look at the frequencies of certain words that spammers tend to write. There's no one word that all spam messages contain that never appears in real messages, but we can nevertheless classify spam very accurately by looking at many different words that all seem to be spam-like.

So what if we could build a "spam filter" for cancer cells? As with suspicious words in an e-mail, it might be hard to find one suspicious molecule, A, that is present in cancer cells and *only* cancer cells. But by also looking at suspicious molecules

B, C, and D, we might be able to do much better at telling the cancer cells apart from normal cells. This might allow us to kill cancer cells with much higher specificity—and that would mean more effective treatments with fewer side effects.

Developing a "signature" of cancer sounds great, but checking each cell individually for this cancer signature could be difficult. Here's a wild idea out of professor of biological engineering Ron Weiss's lab at MIT[15]: maybe we could program all of the cells in the body to examine themselves, then commit suicide if they look like a cancer cell. Using some of the synthetic biology methods discussed in Chapter 9, scientists might someday be able to design a set of genes that, when inserted into a cell, check for various kinds of molecules that might indicate cancer. By measuring perhaps a half dozen molecules that either indicate or rule out cancer, the system encoded by those inserted genes could decide if this cell is cancerous or not. If it's inside a cancer cell, the program triggers a cell suicide pathway, but it doesn't harm noncancerous cells.

We're probably still far away from programming cancer cells to kill themselves. But what if we could specifically attack the cancer cells by taking advantage of a system that's already in the body?

I hadn't necessarily anticipated that I'd be talking about how we might be able to make a vaccine for cancer in a restaurant called Lone Star Taco Bar. But that's where Kelly Moynihan, a graduate student of biological engineering at the Massachusetts Institute of Technology, wanted to meet.

In her research, Moynihan is working on a way we might be able to treat cancer. Her motivations are pretty personal: "My dad was diagnosed with renal cell carcinoma when I was a senior in high school," Moynihan says. "When he was diagnosed, it had already metastasized to his lungs and to his brain, and the five-year survival rate of that type of cancer was under five percent." But against all odds, he survived. He was admitted to a clinical trial for a then-experimental drug called pazopanib, and he got better. Many years later, he's still in remission.

Still, it's not lost on Moynihan how long those odds were. When she went home for Thanksgiving in 2014, she found out that a couple her aunt and uncle know also had come up against her dad's type of cancer. "The husband was diagnosed with the same type of cancer, same stage as my dad's," she says, "and he was dead in three months."

Moynihan's personal experience with cancer drives her to figure out ways to use the immune system against it. Your immune system is already "exquisitely tuned to be able to pick out things that are not you," she explains, so anything that the immune system can recognize as foreign gets attacked and destroyed. "The immune system's already been evolved for millions and millions of years to do this," Moynihan says. "If we have that as a tool, why shouldn't we use it?"

The problem is that the cancers that grow large enough for us to detect are probably either those that have figured out how to evade the immune system or those that the immune system doesn't really recognize as being a problem. Since cancer is made of the patient's own cells, the immune system doesn't always target it effectively—cancer isn't really foreign, at least initially.

Luckily, we already know how to teach the immune system to attack something; we do it all the time with vaccines. Vaccines for diseases like the flu or measles are just dead or nonfunctional virus bits. By injecting these bits into you, we're showing them to your immune system and training it to recognize those bits as bad guys.

So what if we could make a cancer vaccine—to train your immune system to see your cancer as the enemy? This is a little different from the HPV vaccine mentioned in Chapter 12. The HPV vaccine trains your immune system to kill the HPV virus, which can mess up cells in a way that sometimes leads to cancer. Since the vaccine can only prevent

HPV from taking hold in an uninfected person, it's not useful once the patient already has cervical cancer. In contrast, a patient would get a "cancer shot" after they already have the disease, and the shot would teach the immune system to attack the cancer directly. After a cancer diagnosis, the patient would get a vaccine that teaches the immune system that the patient's specific type of cancer is an invader.[16] "The dream is if you could educate the immune system to treat cancers like they do acute viral infections," says Moynihan.

The idea of using the immune system as a weapon against cancer actually isn't a new one. In the 1890s, William Coley, a surgeon at Memorial Hospital in New York, injected his cancer patients with bacteria to stimulate their immune systems to fight the cancer.[17] He was convinced his treatments worked, but his contemporaries were skeptical, and his approach never really caught on. These days, playing around with the immune system makes many scientists nervous precisely because cancer looks similar to healthy human cells; if something goes wrong, there's the danger of accidentally turning the immune system against the patient's own body, usually with nasty results.

Moynihan's current project involves attempting to treat cancer-stricken mice using an approach called combination immunotherapy. "There are all these different types of therapeutics that impact different arms of the immune system," she says. "The immune system is so connected as a network," she explains, so she and her lab wonder: could you stimulate it in different places to get synergistic effects? Can manipulating this system in multiple places at once help boost its tumor-fighting potential?

At least in mice, they see amazingly positive results. We've cured cancer in mice many times before, and those breakthroughs don't always translate to people. But for now, things are looking promising. Moynihan is a member of Massachusetts Institute of Technology professor of materials science and engineering and biological engineering Darrell Irvine's lab, and their group is trying to streamline the different parts of their combination therapy and figure out what each one is doing and why they seem to work well together. She hopes they can use this as a tool to help learn something about the immune system, too.

Even if a cancer vaccine doesn't end up being an effective treatment, there are other ways we might turn the immune system against a tumor. One way is taking immune cells out of the body, "training" them to recognize the cancer in the lab, and putting them back into the patient. Moynihan and I were discussing one treatment that uses this strategy that looks quite promising but will probably be *really* expensive—likely several hundreds of thousands of dollars, in part because of the hands-on time it will require from a scientist to prepare the cells individually for each patient. There's another treatment already on the market that requires similar treatment of immune cells outside of the body, but it's not exactly a home run—it costs about $100,000 per patient and extends lifespan by just four months on average.

But there's one way we can use the immune system against cancer that's already being implemented to great effect. In addition to looking a lot like your healthy cells, cancer cells try to actively suppress the immune system. A new class of drugs called checkpoint inhibitors coming on the market try to prevent this immune suppression. One of them, the not-so-memorably-named pembrolizumab, is used to block one of the methods that melanoma can use to suppress the immune response. The drug seems to be quite successful so far; when former United States president Jimmy Carter was diagnosed with advanced metastatic melanoma in the summer of 2015, doctors treated him with pembrolizumab and he was in remission by Christmas. Only time will tell if President Carter's remission proves long-lasting, but his initially promising response to the treatment is emblematic of the high hopes many scientists have for drugs like these. With any luck, more cancer immunotherapy treatments will be close behind.

Epilogue

Right outside a busy conference room on the fifth floor of the Warren Alpert Building at Harvard Medical School hangs a single, framed page of text. I've never seen anyone stop to read it; perhaps everyone else has already done so. That page is an editorial that was written in the fall of 1960 by James Bonner, then a scientist at Caltech. In this article, "Thoughts on Biology," Bonner lays out what he sees as the future for his field. He describes a class of problems that he calls "systems biology" in which the challenge is "how to use the various and ingenious molecular devices invented by creatures to make a creature or a society."

Even in 1960, it was clear to Bonner that biologists who wanted to tackle these challenges were going to need more training in math and other sciences. "Biology is becoming a rigorous science with sophisticated laws and operational rules," he wrote. "And it appears clear to me that the great forward steps in these complicated problems of biological systems are being made not by formal biologists but by mathematicians, physicists, and even engineers whose interests have turned (probably for some sinister and subconscious reason) to biology." And though I'm not sure I would characterize the motivations of the scientists from other disciplines who are coming to biology as "sinister," Bonner is right on in this respect. We have needed tools from every branch of science in order to make progress on the toughest biological systems. Finally, Bonner gives some career advice: "It should be clear to any promising young person who wishes to solve complex problems of biology that he should start out by becoming, say, a physicist or a mathematician."

More than a half century later, the discipline that Bonner and others like him had envisioned has come to fruition. We understand that a great number of the problems we care most about in biology today are the result of systems, and we're making considerable headway toward understanding those systems. There are many peculiar phenomena in biology, but what is most amazing is that we now understand how some of these peculiar things happen. These behaviors are the product of complex, living systems that are difficult to study. But with a lot of work and some help from new technologies, scientists are beginning to understand and control these systems.

Still, our basic understanding of life will continue to improve. To that end, we will need to continue to develop those new technologies that let us measure all of the important parts of the cell accurately at a scale that enables us to really understand how everything fits together. And with the ever-increasing power of computers, we'll have the firepower we need to ask really big questions: How do cells make decisions? Respond to their environments? To put it simply: How does life work?

The discoveries that we make in the course of this research will have an enormous impact on our lives: having a better understanding of what changes in the microbiome are causing symptoms in people should help us design interventions to prod the system into beneficial states; a deeper understanding of the systems biology of the brain could lead to better treatments for Alzheimer's, Parkinson's, depression, and other brain-related illnesses; and cancer immunotherapy and personalized medicine promise to change the way we treat some complex diseases.

But ultimately, for many systems biologists, future progress will be driven not only by a desire to help people, but also by a burning curiosity about the world. Andrew Murray, a co-director of the systems biology Ph.D. program at Harvard, calls this motivation a desire "to see the face of God." Indeed, learning how these systems work pulls back the curtain on the inner workings of nature and gives us a new appreciation for the living organism. As Terry Pratchett writes in his wonderful children's book *A Hat Full of Sky*, "It's still magic even if you know how it's done."

Acknowledgments

I am indebted to Rachel Vogel for picking me out of the slush pile and for convincing me that systems biology might be a cool thing to write about. Thanks to the team at Bloomsbury for taking a chance on this story; I owe much to Jackie Johnson and her disciplined edits for making this book at least moderately readable. To everyone who read the manuscript or took the time out of their infinitely busy schedules to grant me an interview, thank you for helping to bring these stories to life. My parents also deserve all the credit in the world—not just for letting me do science fair experiments in our basement, but also because I would never have thought to write any of this down if it hadn't been for the need to explain to them what I do for a living. Thanks to Justin, Travis, and sweet baby Griffin for making sure this book wasn't finished too quickly. Thanks finally to the 2004 New York Yankees for raising public awareness of the consequences of not knowing the Heimlich maneuver.

Notes

Preface: The Big Idea

1 Bianconi, E., et al. "An estimation of the number of cells in the human body." *Ann. Hum. Biol.* 40, 463–71 (2013).
2 Lodish, H., Berk, A., and Zipursky, S., in *Molecular Cell Biology* (W. H. Freeman, 2000). At http://www.ncbi.nlm.nih.gov/books/NBK21473/. Accessed August 17, 2016.
3 I estimated the volume of Fenway Park as a rectangular prism, 500 feet on a side and 37 feet high, the height of the Green Monster. A one-inch gumball has a volume of 0.52 cubic inches, and the densest possible sphere packing occupies 74% of the volume of a given space. This gives an estimate of 1,771 Fenway Park volumes, which I round down to 1,000 to be conservative.
4 When I checked in December 2015, I could get 257 gumballs from Amazon.com for $12.76, so 40 trillion gumballs would cost me about $1.986 trillion, assuming I couldn't negotiate a bulk discount. According to the United Nations Statistics Division, Russia's 2013 GDP was $2.097 trillion.

Chapter 1: Seeing the Systems in Biology

1 For more information on Wieschaus and Nüsslein-Volhard's work, see the award ceremony speech announcing the 1995 Nobel Prize in Physiology or Medicine. All of these speeches are archived on the Nobel Committee's website, www.nobelprize.org.

Chapter 2: Déjà Vu All Over Again

1 Spoiler alert: if you've seen the movie, you'll know that one of the apparently good characters is actually a robot

the whole time. For the record, this twist makes exactly as much sense when you can't understand one iota of the dialogue.

2 For our purposes, "network" and "system" are essentially synonyms, but many scientists might draw a distinction: a network is any group of components that are connected in a defined way; a system can be thought of as a type of network where the pieces work together to perform some task or produce some coherent behavior. One might also choose the word "network" over "system" when the connections between components aren't very "real"—for example, a friendship between two people might belong in a network, but perhaps not in a system.

3 Milo, R., et al. "Network motifs: simple building blocks of complex networks." *Science* 298, 824–27 (2002).

4 The full name is β-galactosidase, but ink is expensive. "β-gal" is the commonly used shorthand; it is pronounced "beta-gal," where "gal" rhymes with the "al" in "California."

5 Actually, it grabs on to allolactose, which is made from lactose. But you can think of the presence of allolactose as a good indicator that there's lactose around.

6 This works because the repressor "wins" when both branches are turned on.

7 The "pause" is still under study; it might not be fair to say that the cells are waiting to turn on lactose-digesting genes, so I've simply stated the order of events.

8 This is a simplification, of course.

9 Since proteins are made using RNA as a template instead of DNA, they're actually using another base, U, instead of T. But U behaves almost exactly like T, so it's safe to ignore the difference for the purposes of this book.

10 A very rough calculation: assume that the correct and incorrect matches differ by one hydrogen bond, and assume that one hydrogen bond represents an improvement of about 5 to 10 kJ/mol of free energy; see Turner, et al. *J. Am. Chem. Soc.* 109, 3783–85 (1987), for one measurement that lands in this approximate range. The

ratio of correct to incorrect matches at equilibrium is approximately $e^{\Delta/kT}$, where Δ is the free energy of a mole of hydrogen bonds, k is the Boltzmann constant, and T is the temperature in Kelvin (310 K assumed). If the bond is 5 kJ/mol, this calculation suggests that the correct match will happen only 7 times as often as an incorrect match; 10 kJ/mol implies that correct matches are 50 times as abundant as incorrect matches. Either way, this is well below the accuracy needed to make a protein correctly!

11 "What is the error rate in transcription and translation?" BioNumbers. At http://www.book.bionumbers.org/what-is-the-error-rate-in-transcription-and-translation/. Accessed June 23, 2016.

12 Merlin, C., Gegear, R., and Reppert, S. "Antennal circadian clocks coordinate sun compass orientation in migratory monarch butterflies." *Science* 325, 1700–04 (2009).

13 Delezie, J., et al. "The nuclear receptor REV-ERBα is required for the daily balance of carbohydrate and lipid metabolism." *FASEB J.* 26, 3321–35 (2012).

14 Nakajima, M., et al. "Reconstitution of circadian oscillation of cyanobacterial KaiC phosphorylation in vitro." *Science* 308, 414–15 (2005).

15 In this case, evolution may accept this decreased efficiency in exchange for better performance at extreme temperatures. That shouldn't stop you from feeling superior, though.

Chapter 3: America's Next Top Mathematical Model

1 The equation was so long largely because Gunawardena and his colleagues were working with a system away from equilibrium in which the calculations were history-dependent. The full equation can be found in the supplementary material of Ahsendorf, T., Wong, F., Eils, R., and Gunawardena, J. "A framework for modelling

gene regulation which accommodates non-equilibrium mechanisms." *BMC Biol.* 12, 102 (2014).

2 Inoué, S. "Cell division and the mitotic spindle." *J. Cell Biol.* 91, 131s–47s (1981).

3 Brugués, J., and Needleman, D. "Physical basis of spindle self-organization." *Proc. Natl. Acad. Sci.* 111, 18496–500 (2014).

4 Actually, the mixing seen in coffee is mostly a result of convection, not diffusion. I am glossing over the distinction here because differences between the two processes are not important for the purposes of this book. The necessary intuition—that diffusion causes a "spreading out"—is preserved.

5 See Sheth, R., et al. "Hox genes regulate digit patterning by controlling the wavelength of a Turing-type mechanism," *Science* 338, 1476–80 (2012); and Miura, T., "Turing and Wolpert work together during limb development," *Sci. Signal.* 6, pe14 (2013), for a good nontechnical explanation.

6 Panas, D., et al. "Homeostasis in large networks of neurons through the Ising model: Do higher order interactions matter?" *BMC Neurosci.* 14, 166 (2013).

7 See Dourlent, M., "Competitive cooperative bindings of a small ligand to a linear homopolymer. I. Extension of the Ising model to the case of two competitive interactions," *Biopolymers* 14, 1717–38 (1975); and Liu, Y., and Dilger, J. P. "Application of the one- and two-dimensional Ising models to studies of cooperativity between ion channels," *Biophys. J.* 64, 26–35 (1993).

8 Lee, E. D., Broedersz, C. P., and Bialek, W. "Statistical mechanics of the US Supreme Court." *J Stat Phys* 160, 275–301 (2015).

9 For an interesting justification for this type of endeavor, see Carrera, J. and Covert, M. W., "Why build whole-cell models?" *Trends Cell Biol.* 25, 719–22 (2015).

10 InvivoGen. "Mycoplasma: the insidious invader of cell cultures" (2005). At http://www.invivogen.com/docs/Insight200511.pdf. Accessed August 17, 2016.

11 Karr, J. R., et al. "A whole-cell computational model predicts phenotype from genotype." *Cell* 150, 389–401 (2012).
12 Sanghvi, J. C., et al. "Accelerated discovery via a whole-cell model." *Nat. Methods* 10, 1192–95 (2013).
13 As recounted in Dyson, F., "A meeting with Enrico Fermi." *Nature* 427, 297 (2004). Some scientists have taken this challenge quite literally: Mayer, J., Khairy, K., and Howard, J. "Drawing an elephant with four complex parameters," *Am. J. Phys.* 78, 648–49 (2010).
14 Machta, B. B., et al. "Parameter space compression underlies emergent theories and predictive models." *Science* 342, 604–07 (2013).
15 See Brown, K. S., and Sethna, J. P. "Statistical mechanical approaches to models with many poorly known parameters," *Phys. Rev. E.* 68, 021904 (2003); and Gutenkunst, R. N., et al. "Extracting falsifiable predictions from sloppy models," *Ann. N. Y. Acad. Sci.* 1115, 203–11 (2007).

Chapter 4: Ignoring the Devil in the Details

1 See Barkai, N., and Leibler, S. "Robustness in simple biochemical networks." *Nature* 387, 913–17 (1997) and Alon, U., Surette, M. G., Barkai, N., and Leibler, S., "Robustness in bacterial chemotaxis." *Nature* 397, 168–71 (1999).
2 Some patients must draw a small amount of blood every time they want to measure their glucose levels. Others use implanted measuring devices that provide a continuous readout of glucose levels, but these devices still require regular calibration from a blood sample.
3 Doyle, F. J., et al. "Closed-loop artificial pancreas systems: Engineering the algorithms." *Diabetes Care* 37, 1191–97 (2014).
4 See Dassau, E., et al. "Clinical evaluation of a personalized artificial pancreas." *Diabetes Care* 36, 801–09 (2013) and Dassau, E., et al. "Adjustment of open-loop settings to improve closed-loop results in type 1 diabetes: A

multicenter randomized trial." *J. Clin. Endocrinol. Metab.* 100, 3878–86 (2015).

5 For example, see Teff, K. L., Mattes, R. D., and Engelman, K., "Cephalic phase insulin release in normal weight males: verification and reliability," *Am. J. Physiol.* 261, E430–36 (1991); and Teff, K. L., and Townsend, R. R., "Early phase insulin infusion and muscarinic blockade in obese and lean subjects," *Am. J. Physiol. Regul. Integr. Comp. Physiol.* 277, R198–208 (1999).

6 Most devices actually measure glucose in tissues rather than directly from the blood. This adds another layer of complication and delay to the process: the glucose levels in tissues lag a bit behind the levels in the blood.

7 Lee, J. B., et al. "Model-based personalization scheme of an artificial pancreas for type 1 diabetes applications." *Am. Control Conf.* 2911–16 (2013).

8 Pagliuca, F. W., et al. "Generation of functional human pancreatic β cells in vitro." *Cell* 159, 428–39 (2014).

9 Mitchell, A., et al. "Adaptive prediction of environmental changes by microorganisms." *Nature* 460, 220–24 (2009).

10 Since gene transcription tends to occur in "bursts"—that is, if a cell makes one copy of the gene, it's much more likely to make a second, or third, or tenth—the noise in a cell is probably even worse than it is when knocking pennies off tables. It's as if the pennies clumped together and tended to fall as a group: if one falls, it's bringing others with it.

11 Lestas, I., Vinnicombe, G., and Paulsson, J. "Fundamental limits on the suppression of molecular fluctuations." *Nature* 467, 174–78 (2010).

12 Take x_1 to be the abundance of the mRNA; σ_1 is the standard deviation of the mRNA; and N_1 and N_2 are the number of mRNAs and proteins, respectively, created on average during a time τ. Then the actual bound is $\frac{\sigma_1^2}{\langle x_1 \rangle^2} \geq \frac{1}{\langle x_1 \rangle} \frac{2}{1+\sqrt{1+4N_2/N_1}}$, or approximately $1/\sqrt{N_1 N_2}$

when $N_2 > N_1$. See Lestas, et al. *Nature* 467, 174–78 (2010) for full details.

13 Gregor, T., et al. "Probing the limits to positional information." *Cell* 130, 153–64 (2007).

14 Balaban, N. Q., et al. "Bacterial persistence as a phenotypic switch." *Science* 305, 1622–25 (2004).

15 Kussell, E., and Leibler, S. "Phenotypic diversity, population growth, and information in fluctuating environments." *Science* 309, 2075–78 (2005).

16 Lewis, K. "Persister cells." *Annu. Rev. Microbiol.* 64, 357–72 (2010).

17 For one example of many, see Minkel, J., "Focus: the computer minds the commuter." *Phys. Rev. Focus* 13, 26 (2004).

18 These models also give us hints on how to drive that might minimize traffic jams. If a small number of cars on the road try to drive at the same average speed and "smooth out" stop-and-go traffic, these sorts of traffic jams no longer happen; there have been some reports that cruise control usage, for example, cuts down on traffic.

19 Gordon, D. M., Holmes, S., and Nacu, S. "The short-term regulation of foraging in harvester ants." *Behav. Ecol.* 19, 217–22 (2008).

20 Gordon, D. M. "The rewards of restraint in the collective regulation of foraging by harvester ant colonies." *Nature* 498, 91–93 (2013).

21 Most ants can't see, so they can't literally watch other ants return from foraging; instead, one ant touches the other with its antennae and notices it via olfaction.

22 Gordon, D. M. "The dynamics of foraging trails in the tropical arboreal ant *Cephalotes goniodontus*." *PLoS One* 7, e50472 (2012).

23 Prabhakar, B., Dektar, K. N., and Gordon, D. M. "The regulation of ant colony foraging activity without spatial information." *PLoS Comput. Biol.* 8, e1002670 (2012).

24 The work I'm most familiar with is that of Radhika Nagpal at Harvard; see Werfel, J., Petersen, K., and Nagpal, R.,

"Designing collective behavior in a termite-inspired robot construction team." *Science* 343, 754–58 (2014); and Rubenstein, M., Cornejo, A., and Nagpal, R. "Programmable self-assembly in a thousand-robot swarm." (2014). *Science* 345, 795–99.

Chapter 5: Beyond Tom Hanks's Nose

1 Wilson, M. R., et al. "Actionable diagnosis of neuroleptospirosis by next-generation sequencing." *N. Engl. J. Med.* 370, 2408–17 (2014).

2 They also looked at the RNA in the cerebrospinal fluid, but since the information in RNA is more or less just a copy of the information in the DNA, we'll just talk about DNA for the moment. RNA will be introduced shortly.

3 Technically, Moore's law is usually framed in terms of a doubling of transistor density, but the idea is close enough that no meaning is lost here.

4 An iPhone 6s uses an A9 chip clocked at around 1.8 GHz, according to online benchmarking websites; the Apollo 11 guidance computer was clocked at around 0.043MHz. See Saran, C. "Apollo 11: the computers that put man on the moon." *Computer Weekly* (2009). At http://www.computerweekly.com/feature/Apollo-11-The-computers-that-put-man-on-the-moon. Accessed August 17, 2016.

5 Some viruses use only RNA, but it's debatable whether viruses are "alive"; viruses rely on hijacking living cells and reprogramming them in order to reproduce. Without a cell to take over, an RNA virus is just an inert string of RNA in a protective coating. In any case, all living things use the genetic code (A, G, C, and T) written in DNA and/or RNA to carry information from one generation to another.

6 It's worth noting that there were strategies for measuring a bunch of genes at once that existed before RNA sequencing, such as something called a "microarray." For a review of

the technology from 2006, when microarrays were in their heyday, check out Hoheisel, J. D., "Microarray technology: beyond transcript profiling and genotype analysis." *Nat. Revs. Genet.* 7, 200–210 (2006). doi:10.1038/nrg1809.

7 The method has nothing to do with the direction. It was named by analogy to the Southern blot, a similar technique for detecting DNA instead of RNA, that was invented by the biologist Edwin Southern.

8 X-ray film works using the same principles as the film one might use in a camera. For more on Northern blots, see Alwine, J. C., Kemp, D. J., and Stark, G. R., "Method for detection of specific RNAs in agarose gels by transfer to diazobenzyloxymethyl-paper and hybridization with DNA probes." *Proc. Natl. Acad. Sci. U.S.A.* 74, 5350–54 (1977).

9 This is paraphrased from an analogy used by Professor Patrick Brown of Stanford University, as recalled in an interview by Professor Audrey Gasch of the University of Wisconsin.

10 Gasch, A. P., et al. "Genomic expression programs in the response of yeast cells to environmental changes." *Mol. Biol. Cell* 11, 4241–57 (2000).

11 Chasman, D., et al. "Pathway connectivity and signaling coordination in the yeast stress-activated signaling network." *Mol. Syst. Biol.* 10, 759 (2014).

12 See Perou, C. M., et al., "Molecular portraits of human breast tumours." *Nature* 406, 747–52 (2000); and Sørlie, T., et al., "Gene expression patterns of breast carcinomas distinguish tumor subclasses with clinical implications." *Proc. Natl. Acad. Sci. U.S.A.* 98, 10869–74 (2001).

13 Do not take this as medical advice! If you have any medical concerns, please talk to your doctor. He or she will have the most up-to-date information to help guide your treatment. If you're interested in the scientific literature on the topic, check out Hoadley, K. A., et al. "Breast cancer intrinsic subtypes," *Nat. Rev. Clin. Oncol.* (2014), for a review.

14 Paschka, P., et al. "IDH1 and IDH2 mutations are frequent genetic alterations in acute myeloid leukemia and confer adverse prognosis in cytogenetically normal acute myeloid leukemia with NPM1 mutation without FLT3 internal tandem duplication." *J. Clin. Oncol.* 28, 3636–43 (2010).

15 Except for rare cases when two embryos fuse together very early in development. But that's immaterial to the point at hand.

16 Right after fertilization through the first few cell divisions, the cells of the developing embryo are technically called totipotent. Some of these totipotent cells will give rise to part of the placenta, and the others will become pluripotent stem cells. These pluripotent stem cells then go on to make every type of cell in the body. So there are some cell types—the placental cells and other extra-embryonic tissues—that pluripotent stem cells can't make.

17 Takahashi, K. and Yamanaka, S. "Induction of pluripotent stem cells from mouse embryonic and adult fibroblast cultures by defined factors." *Cell* 126, 663–76 (2006).

18 Pang, Z., et al. "Induction of human neuronal cells by defined transcription factors." *Nature* 476, 220–23 (2011).

19 And platelets.

20 Fountain, H. "A first: organs tailor-made with body's own cells." *New York Times* (2012). At http://www.nytimes.com/2012/09/16/health/research/scientists-make-progress-in-tailor-made-organs.html. Accessed August 17, 2016.

21 Takasato, M., et al. "Kidney organoids from human iPS cells contain multiple lineages and model human nephrogenesis." *Nature* 526, 564–68 (2015).

22 Song, J. J., et al. "Regeneration and experimental orthotopic transplantation of a bioengineered kidney." *Nat. Med.* 19, 646–51 (2013).

23 Costanzo, M., et al. "The genetic landscape of a cell." *Science* 327, 425–32 (2010).

24 I created this web using Cytoscape (http://www.cytoscape.org); each edge corresponds to a Pearson

correlation between two genes of absolute value greater than 0.2. I used genetic interaction data from the supplementary material of Costanzo, et al., *Science* 327, 425–32 (2010); these data are publicly available from the authors at http://www.drygin.ccbr.utoronto.ca/~costanzo2009. I retrieved lists of genes with a given function by querying YeastMine (http://www.yeastmine.yeastgenome.org) for all yeast genes with the corresponding gene ontology ID. I drew only the largest connected component, though there are a few other small components.

25 I just colored ORFs annotated with GO:0006281 (DNA repair), for the purposes of illustration.

26 Blomen, V. A., et al. "Gene essentiality and synthetic lethality in haploid human cells." *Science* 350, 1092–96 (2015).

27 There's no data service in the desert, so he'll have to make do with a voice call.

28 In reality, your DNA isn't just one long strand; it's actually separated into chromosomes, so there are multiple strands. This doesn't really matter, though, since the distance measurement we're going to use will answer "very far away" when we ask for the distance between landmarks on two different chromosomes.

29 For an example of the press coverage, see Knox, R. "Genome maps solve medical mystery for California twins." NPR (2011). At http://www.npr.org/sections/health-shots/2011/06/18/137204964/genome-maps-solve-medical-mystery-for-calif-twins. Accessed August 17, 2016. For a scientific paper describing the case, see Bainbridge, M. N., et al. "Whole-genome sequencing for optimized patient managment." *Sci. Transl. Med.* 3, 87re3 (2011).

30 Brunton, S. L., Proctor, J. L., and Kutz, J. N. "Discovering governing equations from data: sparse identification of nonlinear dynamical systems." arXiv:1509.03580 [math.DS].

31 Of course, Newton's laws aren't the end of the story, either! Einstein's general relativity is an even more fundamental description of the underlying rules that govern the

universe. Even relativity is incomplete: its incompatibility with quantum mechanics tells us that there is a still-more-fundamental description of how the universe works that is waiting to be discovered.

Chapter 6: The Smells of the Father

1 The rare male calico cat happens when a cellular mistake gives the cat two X chromosomes *and* a Y chromosome. In this case, the Y chromosome makes the cat develop like a male, but one of the X chromosomes is still inactivated like it is in females.

2 Engreitz, J. M., et al. "The Xist lncRNA exploits three-dimensional genome architecture to spread across the X chromosome." *Science* 341, 1237973 (2013).

3 Rasmussen, T. P., et al. "Expression of Xist RNA is sufficient to initiate macrochromatin body formation." *Chromosoma* 110, 411–20 (2001); and Hall, L. L., et al. "An ectopic human XIST gene can induce chromosome inactivation in postdifferentiation human HT-1080 cells." *Proc. Natl. Acad. Sci. U.S.A.* 99, 8677–82 (2002).

4 Technically, *Xist* is a gene—it just doesn't happen to make a protein. The "gene" is just the unit of inheritance. I won't stress this point, though, since most genes code for proteins, and functional RNAs are different enough to merit discussion without introducing confusion about what a gene really is.

5 Pray, L. A. "Eukaryotic genome complexity." *Nat. Educ.* 1, 96 (2008).

6 Strachan, T., and Read, A. *Human Molecular Genetics.* (Wiley-Liss, 1999).

7 *NOVA scienceNOW.* "RNAi." (2005). At http://www.pbs.org/wgbh/nova/body/rnai.html. Accessed August 17, 2016.

8 "Rich Jorgensen discusses co-suppression, his work, and the future." *RNAi News* (2004). At https://www.genomeweb.com/rnai/rich-jorgensen-discusses-co-suppression-his-work-and-future. Accessed August 17, 2016.

9 Li, Y., and Kowdley, K. V. "MicroRNAs in common human diseases." *Genomics Proteomics Bioinformatics* 10, 246–53 (2012).
10 Fang, Y., et al. "MicroRNA-7 inhibits tumor growth and metastasis by targeting the phosphoinositide 3-kinase/Akt pathway in hepatocellular carcinoma." *Hepatology* 55, 1852–62 (2012).
11 Calin, G. A., et al. "Frequent deletions and down-regulation of micro-RNA genes *miR15* and *miR16* at 13q14 in chronic lymphocytic leukemia." *Proc. Natl. Acad. Sci. U.S.A.* 99, 15524–29 (2002).
12 See Coelho, T., et al. "Safety and efficacy of RNAi therapy for transthyretin amyloidosis." *N. Engl. J. Med.* 369, 819–29 (2013); and Garber, K. "Alnylam's RNAi therapy targets amyloid disease." *Nat. Biotechnol.* 33, 577 (2015).
13 Lynch, K. W., and Maniatis, T. "Assembly of specific SR protein complexes on distinct regulatory elements of the *Drosophila doublesex* splicing enhancer." *Genes Dev* 10, 2089–101 (1996).
14 Tazi, J., Bakkour, N., and Stamm, S. "Alternative splicing and disease." *Biochim. Biophys. Acta.* 1792, 14–26 (2009).
15 Wilkie, S. E., et al. "Disease mechanism for retinitis pigmentosa (RP11) caused by missense mutations in the splicing factor gene PRPF31." *Mol. Vis.* 14, 683–90 (2008).
16 Jenkins, L. M. M., et al. "p53 N-terminal phosphorylation: a defining layer of complex regulation." *Carcinogenesis* 33, 1441–49 (2012).
17 Ibid.
18 Ikeda, F., and Dikic, I. "Atypical ubiquitin chains: new molecular signals." *EMBO Rep.* 9, 536–42 (2008).
19 Teixeira, L. K., and Reed, S. I. "Ubiquitin ligases and cell cycle control." *Annu. Rev. Biochem.* 82, 387–414 (2013).
20 Semenza, G. L. "Hydroxylation of HIF-1: oxygen sensing at the molecular level." *Physiology (Bethesda).* 19, 176–82 (2004).

21 Technically, I'm referring to osmotic stress brought on via sorbitol treatment, but calling it "dehydrating conditions" evokes the right idea without getting caught up in extraneous concepts.

22 Mody, A., Weiner, J., and Ramanathan, S. "Modularity of MAP kinases allows deformation of their signalling pathways." *Nat. Cell Biol.* 11, 484–91 (2009).

23 Angelman, H. "'Puppet' children: a report on three cases." *Dev. Med. Child Neurol.* 7, 681–88 (1965).

24 Ishmael, H. A., Begleiter, M. L., and Butler, M. G. "Drowning as a cause of death in Angelman syndrome." *Am. J. Ment. Retard.* 107, 69–70 (2002).

25 Ong-Abdullah, M., et al. "Loss of *Karma* transposon methylation underlies the mantled somaclonal variant of oil palm." *Nature* 525, 533–37 (2015).

26 World Wildlife Foundation. "Which everyday products contain palm oil?" (2015). At http://www.worldwildlife.org/pages/which-everyday-products-contain-palm-oil. Accessed August 17, 2016.

27 Kubis, S. E., at al. "Retroelements, transposons and methylation status in the genome of oil palm (*Elaeis guineensis*) and the relationship to somaclonal variation." *Plant Mol. Biol.* 52, 69–79 (2003).

28 Mgbeze, G. C., and Iserhienrhien, A. "Somaclonal variation associated with oil palm (*Elaeis guineensis* Jacq.) clonal propagation: a review." *African J. Biotechnol.* 13, 989–97 (2014).

29 Ong-Abdullah, M., et al. "Loss of *Karma* transposon methylation underlies the mantled somaclonal variant of oil palm." *Nature* 525, 533–37 (2015).

30 See "International Human Genome Sequencing Consortium." "Initial sequencing and analysis of the human genome." *Nature* 409, 860–921 (2001) for the fraction of the genome that consists of transposable elements. For more about retrotransposons, see Cordaux, R., and Batzer, M. A. "The impact of retrotransposons on human genome evolution." *Nat. Rev. Genet.* 10, 691–703 (2009).

31 I fear writing the word "epigenetic" because it means many different things to many different people. Here, I am simply referring to some heritable effect that does not involve a change in the DNA sequence.

32 There is plenty of interesting new research on epigenetic inheritance; it's a fast-moving field. A few references for the interested reader include Bale, T. L. "Epigenetic and transgenerational reprogramming of brain development." *Nat. Rev. Neurosci.* 16, 332–44 (2015); Rodgers, A. B., et al. "Paternal stress exposure alters sperm microRNA content and reprograms offspring HPA stress axis regulation." *J. Neurosci.* 33, 9003–12 (2013); Sharma, U., et al. "Biogenesis and function of tRNA fragments during sperm maturation and fertilization in mammals." *Science* 351, 391–96 (2015); and Padmanabhan, N., et al. "Mutation in folate metabolism causes epigenetic instability and transgenerational effects on development." *Cell* 155, 81–93 (2013).

33 We would typically say that true transgenerational inheritance would have to show an effect that is passed to a male's grandchildren and to a female's great-grandchildren—it's one more generation for females since exposing the female could also directly expose her developing fetus and the cells of that fetus that will eventually become eggs themselves.

34 There are also other, stranger mechanisms that could be responsible, such as structural templating. This involves a protein in one shape serving as a template or a mold to induce other copies of that protein to change into that same shape. This is known to happen in yeast, and it's the same principle that is at work in prion diseases, such as "mad cow" disease.

35 Bygren, L. O., Kaati, G., and Edvinsson, S. "Longevity determined by paternal ancestors' nutrition during their slow growth period." *Acta Biotheor.* 49, 53–59 (2001).

36 Kaati, G., Bygren, L. O., and Edvinsson, S. "Cardiovascular and diabetes mortality determined by nutrition during parents' and grandparents' slow growth period." *Eur. J. Hum. Genet.* 10, 682–88 (2002).

37 Pembrey, M. E., et al. "Sex-specific, male-line transgenerational responses in humans." *Eur. J. Hum. Genet.* 14, 159–66 (2006).

Chapter 7: Growing Pains

1 Koschwanez, J. H., Foster, K. R., and Murray, A. W. "Sucrose utilization in budding yeast as a model for the origin of undifferentiated multicellularity." *PLoS Biol.* 9, e1001122 (2011).
2 There are some exceptions. For example, asexual flatworms called Planaria can reproduce by "tail dropping": their tail completely falls off their body and a whole new animal grows from that tail. Organisms that reproduce sexually start off as a single cell—the fertilized egg.
3 Based on a sketch that appears in Lestas, I., Vinnicombe, G., and Paulsson, J. "Fundamental limits on the suppression of molecular fluctuations." *Nature* 467, 174–78 (2010).
4 See Nonaka, S., et al. "De novo formation of left–right asymmetry by posterior tilt of nodal cilia." *PLoS Biol.* 3, e268 (2005); and Yoshiba, S., and Hamada, H. "Roles of cilia, fluid flow, and Ca^{2+} signaling in breaking of left–right symmetry." *Trends Genet.* 30, 10–17 (2014).
5 Yoshiba, S., et al. "Cilia at the node of mouse embryos sense fluid flow for left-right determination via Pkd2." *Science* 338, 226–31 (2012).
6 Townes, P. L., and Holtfreter, J. "Directed movements and selective adhesion of embryonic amphibian cells." *J. Exp. Zool.* 128, 53–120 (1955).
7 Halbleib, J. M., and Nelson, W. J. "Cadherins in development: cell adhesion, sorting, and tissue morphogenesis." *Genes Dev.* 20, 3199–214 (2006).
8 Gilbert, S. *Developmental Biology.* (Sinauer Associates, 2000). At http://www.ncbi.nlm.nih.gov/books/NBK10065/. Accessed August 17, 2016.
9 Ibid.

10 See Sprinzak, D., et al. "Mutual inactivation of Notch receptors and ligands facilitates developmental patterning." *PLoS Comput. Biol.* 7, e1002069 (2011); and Sprinzak, D., et al. "Cis-interactions between Notch and Delta generate mutually exclusive signalling states." *Nature* 465, 86–90 (2010).

11 Blair, S. S. "Wing vein patterning in *Drosophila* and the analysis of intercellular signaling." *Annu. Rev. Cell Dev. Biol.* 23, 293–319 (2007).

12 For examples, see La Coste, A. de, and Freitas, A. A. "Notch signaling: distinct ligands induce specific signals during lymphocyte development and maturation." *Immunol. Lett.* 102, 1–9 (2006); Louvi, A., and Artavanis-Tsakonas, S. "Notch signalling in vertebrate neural development." *Nat. Rev. Neurosci.* 7, 93–102 (2006); Morell, C. M., et al. "Notch signalling beyond liver development: emerging concepts in liver repair and oncogenesis." *Clin. Res. Hepatol. Gastroenterol.* 37, 447–54 (2013); and Miazga, C. M., and McLaughlin, K. A. "Coordinating the timing of cardiac precursor development during gastrulation: a new role for Notch signaling." *Dev. Biol.* 333, 285–96 (2009).

13 Modified from https://commons.wikimedia.org/wiki/File:Spotted-wing_Drosophila_(Drosophila_suzukii)_female_(15195497409).jpg. Original image by Martin Cooper from Ipswich, UK [CC BY 2.0 (http://www.creativecommons.org/licenses/by/2.0)], via Wikimedia Commons. Accessed August 17, 2016.

14 Modified from a public domain image sourced from Wikimedia: Gray20.png. *Wikimedia Commons* at http://www.en.wikipedia.org/wiki/File:Gray20.png. Accessed August 17, 2016.

Chapter 8: No Organism Is an Island

1 Edward, R., and Roper, C. F. E. "Bioluminescent countershading in midwater animals: evidence from living squid." *Science* 191, 1046–48 (1976).

2 Pollitt, E. J. G., et al. "Cooperation, quorum sensing, and evolution of virulence in Staphylococcus aureus." *Infect. Immun.* 82, 1045–51 (2014).
3 Rutherford, S. T., and Bassler, B. L. "Bacterial quorum sensing: its role in virulence and possibilities for its control." *Cold Spring Harb. Perspect. Med.* 2, a012427 (2012).
4 O'Rourke, J. P., et al. "Development of a mimotope vaccine targeting the *Staphylococcus aureus* quorum sensing pathway." *PLoS One* 9, e111198 (2014).
5 Pickett, J. A., and Griffiths, D. C. "Composition of aphid alarm pheromones." *J. Chem. Ecol.* 6, 349–60 (1980).
6 Sherborne, A. L., et al. "The genetic basis of inbreeding avoidance in house mice." *Curr. Biol.* 17, 2061–66 (2007).
7 Feynman, R. P. *Surely You're Joking, Mr. Feynman!* (W. W. Norton, 1985), p. 97.
8 Kidd, K. A., et al. "Collapse of a fish population after exposure to a synthetic estrogen." *Proc. Natl. Acad. Sci. U.S.A.* 104, 8897–901 (2007).
9 Özen, S., and Darcan, S. "Effects of environmental endocrine disruptors on pubertal development." *J. Clin. Res. Pediatr. Endocrinol.* 3, 1–6 (2011).
10 Vetsigian, K., Jajoo, R., and Kishony, R. "Structure and evolution of streptomyces interaction networks in soil and in silico." *PLoS Biol.* 9, e1001184 (2011).
11 Google Maps data is included in accordance with the print permissions outlined at http://www.google.com/permissions/geoguidelines.html. Imagery is copyright 2015 by Digital Globe and map data is copyright 2015 by Google.
12 Technically, one must also account for whether a gerbil is susceptible or not; gerbils can be immune to the disease from past exposure to plague. This chapter will ignore this factor for simplicity, but the curious reader should simply consider any statements about population density to be about the density of susceptible animals.
13 Davis, S., et al. "The abundance threshold for plague as a critical percolation phenomenon." *Nature* 454, 634–37 (2008).

14 This analogy is imperfect because plague can be transmitted between nonadjacent burrows, but those effects can be accounted for.

15 Mack, R. N., et al. "Biotic invasions: causes, epidemiology, global consequences, and control." *Ecol. Appl.* 10, 689–710 (2000).

16 Since humans reproduce relatively slowly, a gene drive would take centuries to spread through the population. There are also technical hurdles that would need to be solved before a human gene drive would be feasible. Even setting aside the (numerous!) ethical issues, human gene drives aren't really feasible using the methods described here.

17 CRISPR stands for clustered regularly interspaced short palindromic repeats, after one of the features of the bacterial "immune system" that the technology is based on.

18 Isaacs, A. T., et al. "Engineered resistance to *Plasmodium falciparum* development in transgenic *Anopheles stephensi.*" *PLoS Pathog.* 7, e1002017 (2011).

19 Gantz, V. M., et al. "Highly efficient Cas9-mediated gene drive for population modification of the malaria vector mosquito *Anopheles stephensi.*" *Proc. Natl. Acad. Sci. U.S.A.* 112, E6736–43 (2015).

20 The entire story of Lotka and Volterra is based on a very helpful book: Kingsland, S. E. *Modeling Nature: Episodes in the History of Population Ecology.* (University of Chicago Press, 1985), 98–111.

21 Ibid.

22 Ibid.

23 For those who have a little math background, the equations are as follows: $\frac{dR}{dt} = aR(t) - bR(t)F(t)$ and $\frac{dF}{dt} = -cF(t) + dR(t)F(t)$. This system orbits a steady state at $\left(\frac{c}{d}, \frac{a}{b}\right)$.

24 May, R. M. "Will a large complex system be stable?" *Nature* 238, 413–14 (1972).

25 Of course, systems can have more than one steady state, and those steady states can be a mix of stable and unstable fixed

points. For the mathematically inclined reader: here, we're specifically considering a linear expansion about a single steady state and looking at the eigenvalues of the Jacobian.

26 For the mathematically inclined reader: May showed the conditions under which large random matrices have only eigenvalues with negative real parts. This restriction on the Jacobian is necessary and sufficient for stability.

27 Haldane, A. G. and May, R. M. "Systemic risk in banking ecosystems." *Nature* 469, 351–55 (2011).

28 Allesina, S., and Tang, S. "Stability criteria for complex ecosystems." *Nature* 483, 205–08 (2012).

29 Bedi, R. "Explosion in rat population causes Indian famine fear." *Irish Times*, June 7, 2006.

30 Foster, P. "Bamboo threatens to bring Indian famine." *Daily Telegraph* (London), October 14, 2004.

31 Ibid.

32 This story of how the bamboo developed its flowering strategy is speculative, like the answers to most "Why?" questions in biology. (Why does this happen? Because it evolved to be that way. We can hypothesize about the selective pressures that shaped this process, but we can never replay history to be sure.)

Chapter 9: Build Me a Buttercup

1 Howard, T. P., et al. "Synthesis of customized petroleum-replica fuel molecules by targeted modification of free fatty acid pools in *Escherichia coli*." *Proc. Natl. Acad. Sci. U.S.A.* 110, 7636–41 (2013).

2 Building the circuit was feasible, but still difficult. "At the time, and for several years afterwards, almost all genetic engineering was based on simply expressing one gene of interest," Elowitz says. "There was little precedent for putting together a circuit requiring four genes."

3 Elowitz, M. B., and Leibler, S. "A synthetic oscillatory network of transcriptional regulators." *Nature* 403, 335–38 (2000).

4 Beal, J., Lu, T., and Weiss, R. "Automatic compilation from high-level biologically-oriented programming language to genetic regulatory networks." *PLoS One* 6, e22490 (2011).
5 Nielsen, A. A. K., et al. "Genetic circuit design automation." *Science.* 352, aac7341 (2016).
6 Johns Hopkins Malaria Research Institute. "About Malaria." At http://www.malaria.jhsph.edu/about-malaria/. Accessed August 17, 2016.
7 Gallup, J., and Sachs, J. "The Economic Burden of Malaria." In *The Intolerable Burden of Malaria: A New Look at the Numbers: Supplement to Volume 64(1) of the American Journal of Tropical Medicine and Hygiene* (eds. Breman, J., Egan, A., and Keusch, G.). American Society of Tropical Medicine and Hygiene, 2001.
8 For a good general overview of this fascinating subject, check out Serjeant, G. R. "One hundred years of sickle cell disease." *Br. J. Haematol.* 151, 425–29 (2010). For more on the protective effects of the variant, see Allison, A. C. "Protection afforded by sickle-cell trait against subtertian malarial infection." *Br. Med. J.* 1, 290–94 (1954); and Ferreira, A., et al. "Sickle hemoglobin confers tolerance to Plasmodium infection." *Cell* 145, 398–409 (2011).
9 Miller, L., and Su, X. "Artemisinin: discovery from the Chinese herbal garden." *Cell* 146, 855–58 (2011).
10 Hung, L., de Vries, P., and Thuy, L. "Single dose artemisinin-mefloquine versus mefloquine alone for uncomplicated falciparum malaria." *Trans. Royal Soc. Trop. Med. Hygiene* 91, 191–94 (1997).
11 Ro, D.-K., et al. "Production of the antimalarial drug precursor artemisinic acid in engineered yeast." *Nature* 440, 940–43 (2006).
12 The yeast do not actually produce artemisinin directly; instead, they produce a closely related chemical that chemists can then modify to produce the final drug.
13 Acott, C. "J S Haldane, J B S Haldane, L Hill and A Siebe: a brief resume of their lives." *SPUMS J.* 29, 161–65 (1999).

14 Mine Safety and Health Administration of the United States Department of Labor. "A Pictorial Walk Through the 20th Century." At http://www.msha.gov/century/canary/canary.asp. Accessed December 15, 2015.
15 See Davidov, Y., et al. "Improved bacterial SOS promoter::lux fusions for genotoxicity detection." *Mutat. Res.* 466, 97–107 (2000); Polyak, B., et al. "Bioluminescent whole cell optical fiber sensor to genotoxicants: system optimization." *Sensors Actuators B:* 74, 18–26 (2001); and Gu, M., Mitchell, R., and Kim, B. "Whole-cell-based biosensors for environmental biomonitoring and application." *Biomanufacturing* 87, 269–305 (2004).
16 For another example of this type of sensor, see Fendyur, A., et al. "Cell-based biosensor to report DNA damage in micro- and nanosystems." *Anal. Chem.* 86, 7598–605 (2014).
17 Malyshev, D. A., et al. "A semi-synthetic organism with an expanded genetic alphabet." *Nature* 509, 385–88 (2014).
18 Gibson, D. G., et al. "Creation of a bacterial cell controlled by a chemically synthesized genome." *Science* 329, 52–56 (2010).

Chapter 10: More Than Just 86 Billion Neurons

1 Parkinson's Deep Brain Stimulation. At https://www.youtube.com/watch?v=mO3C6iTpSGo. Accessed August 17, 2016.
2 Stufflebeam, R. "Neurons, synapses, action potentials, and neurotransmission." *Consortium on Cognitive Science Instruction* (2008). At http://www.mind.ilstu.edu/curriculum/neurons_intro/neurons_intro.php. Accessed August 17, 2016.
3 Based on the sources collected in the Physics Factbook, "Power of a human brain." At http://www.hypertextbook.com/facts/2001/JacquelineLing.shtml. Accessed June 23, 2016.
4 I got: (1) "It was a dark and stormy night . . . no, that's not it." (2) "I'm certain you've heard it before." (3) "What,

again?" These are all clever tricks to avoid the real question. Siri is many things—including easy-to-anthropomorphize—but she's not creative.

5 The method described here is about as far from the most efficient way to solve this problem as possible, but I optimized for simplicity of explanation.

6 Specifically, I was thinking of a perceptron when I designed this exercise.

7 For just one of countless examples of these types of applications, see Deng, L., et al. "Deep neural networks for acoustic modeling in speech recognition." *IEEE Signal Processing Magazine* 82–97 (November 2012).

8 Simonite, T. "Google's brain-inspired software describes what it sees in complex images." *MIT Technology Review* (2014). At http://www.technologyreview.com/news/532666/googles-brain-inspired-software-describes-what-it-sees-in-complex-images/. Accessed August 17, 2016.

9 Vogel, G. "Malaria as lifesaving therapy." *Science* 342, 686 (2013).

10 Hubel, D., and Wiesel, T. "Receptive fields of single neurones in the cat's striate cortex. *J. Physiol.* 574–91 (1959). At http://www.ncbi.nlm.nih.gov/pmc/articles/PMC1363130/. Accessed August 17, 2016.

11 Shuler, M., and Bear, M. "Reward timing in the primary visual cortex." *Science* 311, 1606–10 (2006).

12 Soon, C. S., et al. "Unconscious determinants of free decisions in the human brain." *Nat. Neurosci.* 11, 543–45 (2008).

13 This experiment does depend on accurate reporting of when the decision happened, which is potentially problematic.

14 Schultze-Kraft, M., et al. "The point of no return in vetoing self-initiated movements." *Proc. Natl. Acad. Sci. U.S.A.* 113, 1080–85 (2015).

15 Chen, X., et al. "High-speed spelling with a noninvasive brain–computer interface." *Proc. Natl. Acad. Sci. U.S.A.* 112, E6058–67 (2015).

16 Aflalo, T., et al. "Decoding motor imagery from the posterior parietal cortex of a tetraplegic human." *Science* 348, 906–10 (2015).
17 Vogelstein, J. T., et al. "Discovery of brainwide neural-behavioral maps via multiscale unsupervised structure learning." *Science* 344, 386–92 (2014).
18 Chen, F., Tillberg, P. W., and Boyden, E. S. "Expansion microscopy." *Science.* 347, 543–48 (2015).

Chapter 11: Death and Taxes

1 For one early example, see McCay, C. M., Crowell, M. F., and Maynard, L. A. "The effect of retarded growth upon the length of life span and upon the ultimate body size." *J. Nutr.* 10, 63–79 (1935).
2 Klass, M. R. "Aging in the nematode *Caenorhabditis elegans*: major biological and environmental factors influencing life span." *Mech. Ageing Dev.* 6, 413–29 (1977).
3 Compare Mattison, J. A., et al. "Impact of caloric restriction on health and survival in rhesus monkeys from the NIA study." *Nature* 489, 318–21 (2012) and Colman, R. J., et al. "Caloric restriction delays disease onset and mortality in rhesus monkeys." *Science* 325, 201–04 (2009).
4 Berrington de Gonzalez, A., et al. "Body-mass index and mortality among 1.46 million white adults." *N. Engl. J. Med.* 363, 2211–19 (2010).
5 Klass, M. R. "Aging in the nematode *Caenorhabditis elegans*: major biological and environmental factors influencing life span." *Mech. Ageing Dev.* 6, 413–29 (1977).
6 Lizards that sun themselves on rocks in order to regulate their internal temperature are also sometimes called "cold-blooded," but they are not poikilothermic. A poikilotherm does not attempt to maintain a constant internal body temperature.
7 Xiao, R., et al. "A genetic program promotes *C. elegans* longevity at cold temperatures via a thermosensitive TRP channel." *Cell* 152, 806–17 (2013).

8 Friedman, D. B., and Johnson, T. E. "A mutation in the *age-1* gene in Caenorhabditis elegans lengthens life and reduces hermaphrodite fertility." *Genetics* 118, 75–86 (1988).

9 Kenyon, C., et al. "A *C. elegans* mutant that lives twice as long as wild type." *Nature* 366, 461–64 (1993).

10 Bishop, N. A., and Guarente, L. "Two neurons mediate diet-restriction-induced longevity in *C. elegans*." *Nature* 447, 545–49 (2007).

11 Burkewitz, K., et al. "Neuronal CRTC-1 governs systemic mitochondrial metabolism and lifespan via a catecholamine signal." *Cell* 160, 842–55 (2015).

12 Hughes, A. L., and Gottschling, D. E. "An early age increase in vacuolar pH limits mitochondrial function and lifespan in yeast." *Nature* 492, 261–65 (2012).

13 Tait, S. W. G., and Green, D. R. "Mitochondria and cell signalling." *J. Cell Sci.* 125, 807–15 (2012).

14 Roubertoux, P. L., et al. "Mitochondrial DNA modifies cognition in interaction with the nuclear genome and age in mice." *Nat. Genet.* 35, 65–69 (2003).

15 Hudson, G., et al. "Recent mitochondrial DNA mutations increase the risk of developing common late-onset human diseases." *PLoS Genet.* 10, e1004369 (2014).

16 Greer, E. L., et al. "Members of the H3K4 trimethylation complex regulate lifespan in a germline-dependent manner in *C. elegans*." *Nature* 466, 383–87 (2010).

17 Greer, E. L., et al. "Transgenerational epigenetic inheritance of longevity in *Caenorhabditis elegans*." *Nature* 479, 365–71 (2011).

18 Hayflick, L., and Moorhead, P. S. "The serial cultivation of human diploid cell strains." *Exp. Cell Res.* 25, 585–621 (1961).

19 See Baker, D. J. et al., "Naturally occurring p16$^{\mathrm{Ink4a}}$-positive cells shorten healthy lifespan." *Nature* 530, 184–89 (2016) and Baker, D. J., et al. "Clearance of p16$^{\mathrm{Ink4a}}$-positive senescent cells delays ageing-associated disorders." *Nature* 479, 232–36 (2011).

20 Shomrat, T., and Levin, M. "An automated training paradigm reveals long-term memory in planarians and its persistence through head regeneration." *J. Exp. Biol.* 216, 3799–810 (2013).
21 See Ruckh, J. M., et al. "Rejuvenation of regeneration in the aging central nervous system." *Cell Stem Cell* 10, 96–103 (2012); Katsimpardi, L., et al. "Vascular and neurogenic rejuvenation of the aging mouse brain by young systemic factors." *Science* 344, 630–35 (2014); and Boutet, S. C., et al. "Alternative polyadenylation mediates microRNA regulation of muscle stem cell function." *Cell Stem Cell* 10, 327–36 (2012).

Chapter 12: Your Microbiome and You

1 UNICEF. "The Situation of Women and Children." At http://www.unicef.org/malawi/children.html. Accessed June 23, 2016.
2 Garrett, W. S. "Kwashiorkor and the gut microbiota." *N. Engl. J. Med.* 368, 1746–47 (2013).
3 Smith, M. I., et al. "Gut microbiomes of Malawian twin pairs discordant for kwashiorkor." *Science* 339, 548–54 (2013).
4 Specifically, RUTF produced a transient "maturation" in the microbiome. During the development of healthy children, the scientists observed that the microbiome underwent changes and "matured," but the malnourished children seemed to have an "immature" microbiome that resembled the microbial community seen in younger children.
5 Blanton, L. V. et al. "Gut bacteria that prevent growth impairments transmitted by microbiota from malnourished children." *Science* 351, aad3311 (2016).
6 Sender, R., Fuchs, S., and Milo, R. "Revised estimates for the number of human and bacteria cells in the body." *bioRxiv* 036103 (2016).
7 Some scientists would use the term "microbiota" to describe the collection of microbes and "microbiome" to refer to the total set of genetic material carried by those microbes.

However, in the popular press, "microbiome" seems to have replaced "microbiota." I'll use "microbiome" to mean the collection of microbes in and on the body as it's the meaning readers are most likely to encounter outside of this book.

8 *Fresh Air.* "Bacterial bonanza: microbes keep us alive." NPR (September 15, 2010). At http://www.npr.org/templates/story/story.php?storyId=129862107. Accessed August 17, 2016.

9 Technically, the analogous group for humans is the chordates, but the distinction is not important for our purposes.

10 They use a computational method described in Langille, M. G. et al. "Predictive functional profiling of microbial communities using 16S rRNA marker gene sequences." *Nat. Biotechnol.* 31, 814–21 (2013).

11 Liou, A. P., et al. "Conserved shifts in the gut microbiota due to gastric bypass reduce host weight and adiposity." *Sci. Transl. Med.* 5, 178ra41 (2013).

12 Lessa, F. C. et al. "Burden of *Clostridium difficile* infection in the United States." *N. Engl. J. Med.* 372, 825–34 (2015).

13 Janeway, C. J., Travers, P., and Walport, M. *Immunobiology: The Immune System in Health and Disease* (Garland Science, 2001). At http://www.ncbi.nlm.nih.gov/books/NBK27169/. Accessed August 17, 2016.

14 American Cancer Society. "Can infections cause cancer?" At http://www.cancer.org/cancer/cancercauses/othercarcinogens/infectiousagents/infectiousagentsandcancer/infectious-agents-and-cancer-intro. Accessed June 23, 2016.

15 More accurately, the ancestral rats who were afraid of cats were more likely than their fearless siblings to survive long enough to reproduce and pass that fear of cats on to their children. Rats who avoid being eaten are able to reproduce and take over the population.

16 These effects were first described in Berdoy, M., Webster, J. P. and Macdonald, D. W. "Fatal attraction in rats infected with *Toxoplasma gondii.*" *Proc. Biol. Sci.* 267, 1591–94 (2000); Vyas, A., et al. "Behavioral changes induced by *Toxoplasma* infection of rodents are highly specific to aversion of

cat odors." *Proc. Natl. Acad. Sci. U.S.A.* 104, 6442–47 (2007); and Lamberton, P. H. L., Donnelly, C. A., and Webster, J. P. "Specificity of the *Toxoplasma gondii*-altered behaviour to definitive versus non-definitive host predation risk." *Parasitology* 135, 1143–50 (2008).

17 Ingram, W. M., et al. "Mice infected with low-virulence strains of *Toxoplasma gondii* lose their innate aversion to cat urine, even after extensive parasite clearance." *PLoS One* 8, e75246 (2013).

18 Perhaps by changing some of the "DNA margin notes" discussed in Chapter 6, thereby affecting gene expression in the amygdala; see Hari Dass, S. A. and Vyas, A. "*Toxoplasma gondii* infection reduces predator aversion in rats through epigenetic modulation in the host medial amygdala." *Mol. Ecol.* 23, 6114–22 (2014).

19 Poirotte, C. et al. "Morbid attraction to leopard urine in *Toxoplasma*-infected chimpanzees." *Curr. Biol.* 26, R98–R99 (2016).

20 Bravo, J. A., et al. "Ingestion of Lactobacillus strain regulates emotional behavior and central GABA receptor expression in a mouse via the vagus nerve." *Proc. Natl. Acad. Sci. U.S.A.* 108, 16050–55 (2011).

21 Bercik, P., et al. "The anxiolytic effect of *Bifidobacterium longum* NCC3001 involves vagal pathways for gut-brain communication." *Neurogastroenterol. Motil.* 23, 1132–39 (2011).

22 Rothhammer, V., et al. "Type I interferons and microbial metabolites of tryptophan modulate astrocyte activity and central nervous system inflammation via the aryl hydrocarbon receptor." *Nat. Med.* 22, 586–97 (2016).

Chapter 13: This Is Your System on Drugs

1 Jenkins, A., et al. "Identification and quantitation of alkaloids in coca tea." *Forensic Sci. Int.* 77, 179–89 (1996).

2 Paul, S. M., et al. "How to improve R&D productivity: the pharmaceutical industry's grand challenge." *Nat. Rev. Drug Discov.* 9, 203–14 (2010).

3 Allosteric sites. *Mosby's Medical Dictionary* (2009). At http://www.medical-dictionary.thefreedictionary.com/allosteric+sites. Accessed August 17, 2016.

4 Abbott, K. "The 1904 Olympic marathon may have been the strangest ever." Smithsonian.com (2012). At http://www.smithsonianmag.com/history/the-1904-olympic-marathon-may-have-been-the-strangest-ever-14910747/?no-ist. Accessed August 17, 2016.

5 Wells, H. G. *The Invisible Man* (2004). At http://www.gutenberg.org/files/5230/5230-h/5230-h.htm. Accessed August 17, 2016.

6 There are other similar types of simulations that represent groups of atoms as a single particle; this strategy achieves increased speed at the cost of some accuracy. Such models are called "coarse grained."

7 Anton's molecular dynamics simulations also account for van der Waals interactions and some subtleties that I'm glossing over here. Van der Waals interactions might be described as a rule that says that all else being equal, atoms like to be close to one another but not so close that they bump into one another.

8 Dror, R. O., et al. "Pathway and mechanism of drug binding to G-protein-coupled receptors." *Proc. Natl. Acad. Sci. U.S.A.* 108, 13118–23 (2011).

9 Filmore, D. "It's a GPCR world." *Mod. Drug Discov.* 7, 24–27 (2004).

10 McKenzie, R., et al. "Hepatic failure and lactic acidosis due to fialuridine (FIAU), and investigational nucleoside analogue for chronic Hepatitis B." *N. Engl. J. Med.* 333, 1099–105 (1995).

11 Gambacorti-Passerini, C. "Part I: milestones in personalised medicine: imatinib." *Lancet Oncol.* 9, 600 (2008).

12 Grskovic, M., et al. "Induced pluripotent stem cells: opportunities for disease modelling and drug discovery." *Nat. Rev. Drug Discov.* 10, 915–29 (2011).

13 Hajdu, S. I. "A note from history: landmarks in history of cancer, part 2." *Cancer* 117, 2811–20 (2011).

14 Mraz, M., et al. "Bone marrow stromal cells protect lymphoma B-cells from rituximab-induced apoptosis and targeting integrin α-4-β-1 (VLA-4) with natalizumab can overcome this resistance." *Br. J. Haematol.* 155, 53–64 (2011).

15 Xie, Z., et al. "Multi-input RNAi-based logic circuit for identification of specific cancer cells." *Science* 333, 1307–11 (2011).

16 Some readers might wonder: if cancer cells come from normal cells, how can the immune system recognize them as "foreign" and target them for destruction? Scientists are still investigating what kinds of signals can tell the immune system that a cell is cancerous, but early results seem to point to mutant peptides as one player. Cancer cells often have defects in the cellular machinery that repairs DNA damage, so these cancer cells accumulate changes to their DNA at a faster rate than normal cells do. When genes that are affected by these mutations produce a protein, that altered protein can sometimes be transported to the cell's surface and "displayed" for immune cells to inspect. Since the mutated form of the protein is unlike the version of the protein that the immune system has become accustomed to seeing, it can be recognized as foreign and provoke an immune response.

17 McCarthy, E. F. "The toxins of William B. Coley and the treatment of bone and soft-tissue sarcomas." *Iowa Orthop. J.* 26, 154–58 (2006).

Index

A Note on the Author

James R. Valcourt is pursuing a Ph.D. in systems biology at Harvard University. He is a recipient of the quarter-million-dollar Fannie and John Hertz Foundation Graduate Fellowship and the National Science Foundation Graduate Research Fellowship. He previously worked as a scientist at D. E. Shaw Research in New York City, where he used supercomputer simulations to study the mechanisms by which pharmaceutical drugs work. James graduated magna cum laude from Princeton University with an A.B. in molecular biology and a certificate in quantitative and computational biology; at Princeton, he received the Moses Taylor Pyne Honor Prize, the highest general distinction conferred on an undergraduate. He lives in Somerville, Massachusetts. In his free time, James enjoys backpacking, listening to comedy podcasts, and telling stories about the microbiome at cocktail parties.